Homeland Security Threats, Countermeasures, and Privacy Issues

Homeland Security Threats, Countermeasures, and Privacy Issues

Giorgio Franceschetti

Marina Grossi

Editors

ARTECH HOUSE

BOSTON | LONDON

artechhouse.com

Library of Congress Cataloging-in-Publication Data
A catalog record for this book is available from the U.S. Library of Congress.

British Library Cataloguing in Publication Data
A catalogue record for this book is available from the British Library.

Cover design by Vicki Kane

ISBN 13: 978-1-60807-106-7

685 Canton Street
Norwood, MA 02062

10 9 8 7 6 5 4 3 2 1

Contents

Preface *xiii*

1 The New Vision of the Homeland Security Scenario 1

1.1 Unveiling the New Scenario 1
1.1.1 Tokyo Chemical Sarin Attack 2
1.1.2 Murrach Federal Building Attack 3
1.1.3 Beslan 3
1.1.4 London Bombing 3
1.1.5 Penetration of the U.S. Electric Grid 4
1.1.6 Vandals Cut Phones and Internet Cables 5
1.1.7 Stuxnet Computer Worm 5
1.2 The New Rules and Operational Avenues 6
1.3 The Last Step 9
References 10

Appendix 1A
Team for Research in Ubiquitous Secure Technologies (TRUST) 11
1A.1 Introduction 11
1A.2 Research Overview 12
1A.3 TRUST Research in Secure Financial Infrastructures 17
1A.4 TRUST Research in Secure Health Infrastructures 21
1A.5 TRUST Research in Secure Physical Infrastructures 25
1A.6 Shared Experimental Facilities 28
1A.7 Concluding Remarks 34
Acknowledgments 35
References 35

2 Homeland Security and National Defense in the Twenty-First Century 43

2.1 The New World Scenario 43

2.2 National Defense—Homeland Protection 45

2.3 Network-Enabled Capability 45

2.4 Net-Centric Warfare Doctrine and Technological Challenges 48

2.5 The NEC Implementation in an Archetypal Nation: Italy 51

2.6 Conclusions 54

Selected Bibliography 55

3 Homeland Security and Challenges in Information Systems 57

3.1 ICT and Society, ICT and Control Systems, and the Blurring from Both 57

3.1.1 Chapter Overview 58

3.1.2 Critical Infrastructure and ICT 58

3.1.3 Critical Infrastructure and Control Systems 61

3.1.4 The Mobility Trend 63

3.1.5 Blurring Borders of ICT, General, and Mobile Infrastructure 64

3.1.6 Globalization and Virtualization 65

3.1.7 Need for Actions on the Policy Side: C(I)IP on the International and European Level 65

3.2 The Information Security Challenge: E-Banking as an Example 67

3.2.1 The Challenge to Secure General-Purpose Computers for E-Banking 67

3.2.2 Countermeasures Against E-Banking Fraud 69

3.3 Analysis of Research Needs in Critical Financial Infrastructure: Parsifal 71

3.3.1 Project Overview 72

3.4 Introducing Information Sharing: Key Challenges 76

3.5 The Dilemma of Security-Privacy 79

3.6 Conclusions 82

References 83
Selected Bibliography 83

4 Analysis of Emerging Phenomena in Large Complex Systems 85

4.1 Introduction 85
4.2 The Challenge of Complex Systems 86
4.3 The New Approach of System Engineering 87
4.4 Study Cases 90
4.4.1 The Traffic Congestion Phenomenon in Internet-Like Networks 91
4.4.2 Domino Effect in a Large High-Voltage Electric Distribution Grid 100
4.4.3 Interdependence Analysis in Large Critical Infrastructures 105
4.5 Conclusions 114
Acknowledgments 115
References 115

5 Model-Based Design of Trustworthy Health Information Systems 119

5.1 Introduction 119
5.2 Patient-Centered Clinical Information Management Processes 122
5.2.1 Patient Portals 123
5.2.2 Patient Management Systems 124
5.3 Policy Languages and Policy Composition 125
5.4 Ethical and Legal Analysis 126
5.5 Model-Based Design 128
5.6 Model-Based Experiment Integration 130
5.7 Closing the Loop: Patient-Centered Health Information Management Evaluation 130
References 131

6 Urban Defense Using Mobile Sensor Platforms: Surveillance, Protection, and Privacy 135

6.1 Introduction 135

6.1.1 Urban Surveillance 135
6.1.2 Video Camera Surveillance 136
6.1.3 Fixed Video Camera Limitations and Emergence of Mobile Sensing 140

6.2 MobEyes: Video Surveillance from Vehicular Platforms 142

6.2.1 Vehicular Sensor Networks 142
6.2.2 Related Work 144
6.2.3 MobEyes Architecture 146

6.3 MobEyes Urban Surveillance Examples 148

6.3.1 The Westwood Scenario 148
6.3.2 Data Validation 149
6.3.3 Tracking Suspects Using MobEyes 152

6.4 Secure, Selective Surveillance in Vehicular Networks 154

6.4.1 Vehicular Security 154
6.4.2 New Security Requirements for Urban Surveillance 155
6.4.3 Attribute-Based Encryption: A Primer 156
6.4.4 DABE: The Better Choice in Dynamic Environments 158
6.4.5 Benefits of Dynamic Attribute-Based Encryption 160
6.4.6 DABE in Homeland Defense Applications 162

6.5 Conclusions 163

References 164

Appendix 6A The C-VeT Testbed 167

6A.1 The C-VeT Architecture 167
6A.2 MobiMESH 170

7 Detection and Identification of Dangerous Materials for Airport Security 173

7.1 The Airport Transport Security Problem 173

7.2 Standard Security Solutions and Issues 176

7.3 Innovative Approach to Security 180

7.4 Sensors for a Distributed Dangerous Object Detection/Classification System 184

7.5 Indoor/Outdoor Noncooperative People Tracking 188

7.6 Conclusions 193

Acknowledgments 193

References 194

Appendix 7A List of Airport Security Measure Failures with Worst Consequences 197

8 Privacy Versus Security: A Fight That May Turn into an Alliance 201

8.1 Introduction 201

8.2 Privacy and Security and Safety: Their Definition and Correlation 203

8.3 The Paradigm of Integrated Security and Safety Management 205

8.4 The New Interpretation on the Security Role 207

8.5 The Regulatory Framework and the Open Model of Assurance 209

8.6 The New Model: The Archetypal Approach 211

8.7 Conclusions 215

References 215

9 Privacy-Aware Design for the Monitoring, Control, and Protection of Critical Infrastructure 217

9.1 Introduction 217

9.2 Sensor Networking Technologies for Infrastructure Protection 218

9.3 Example Application Development: SCADA 220

9.4 Privacy: Definitions and the Law 224

9.5 Privacy Issue: The Market for Information 227

9.6 Protecting Privacy: Privacy-Aware Design 228

9.6.1 Provide Full Disclosure of Data Collection 228

9.6.2 Require Consent to Data Collection 229

9.6.3 Minimize the Collection of Personal Data 230

9.6.4 Minimize the Identification of Data with Individuals 230

9.6.5 Minimize and Secure Data Retention 230

9.7 Privacy Problems and Advanced Sensing System: Demand Response Case Study 231
9.7.1 Power Consumption Data Creates a Privacy Issue 231
9.7.2 A Privacy-Aware Demand Response System 233
9.8 Conclusion 235
References 235

10 Military Defense, Civil Defense, and Civil Protection Integration in a Multiscenario Crisis Event 237

10.1 Introduction 237
10.1.1 Major Events 238
10.1.2 Prospects and Problems 239
10.2 Italian National Protection System 240
10.2.1 Risk Assessment 240
10.2.2 Planning Scenarios 240
10.2.3 Security Tasks 240
10.2.4 National Emergency Management System 241
10.2.5 Recovery and Consequence Management 244
10.3 System Criticality 244
10.4 Summit G8 Security Planning 246
10.4.1 Participants 248
10.4.2 Planning and Organization 249
10.4.3 Command and Control Security System 252
10.4.4 Coordination Table 252
10.4.5 Joint Operation Room 252
10.4.6 Technical Management Center 253
10.4.7 The Armed Forces Operation Room 254
10.4.8 The International Operation Room 254
10.5 Conclusions 254
References 255

11 Repel Boarders! 257

11.1 Introduction 257
11.2 Pirates 258
11.3 The Threat 259
11.4 Prepare to Repel Boarders! 261

11.5 Modern Solutions for the Antipiracy Issue 262
11.5.1 The Ship Segment 262
11.5.2 The Shore Segment 262
11.6 A Recently Implemented Antipiracy System 264
11.6.1 The Medusa EO System 264
11.6.2 The Video Surveillance Unit 267
11.6.3 The LYRA Surveillance Radar 268
11.6.4 The Multifunctional Console 269
11.7 The Validating Experiment 270
11.8 Conclusions 271

About the Authors 273

Index 281

Preface

This book is based on the lectures presented at the International Workshop on Homeland Security, sponsored by SELEX Sistemi Integrati, and held at the Casina Vanvitelliana, Bacoli (NA), September 21–25, 2009. The Casina Vanvitelliana (Figure P.1) is a very attractive hunting cottage, located on a small island in the Fusaro Lake, close to the Mediterranean Sea. The Casina was designed by Luigi Vanvitelli, and completed (1782) by his son Carlo under the reign of Ferdinando IV of the Borbone family. At that time, the southern part of Italy (Sicily included) was unified as an independent sovereignty, namely, Regno

Figure P.1 Picture of the Casina Vanvitelliana, located on a small island in the Fusaro Lake.

delle due Sicilie (Reign of two Sicilies), under the Borbone dynasty. Among other important personalities, the Casina hosted the musician composers Wolfgang Amadeus Mozart, Gioacchino Rossini, and the Czar of Russia at that time.

In this magnificent location, helped by beautiful weather, superior catering food, and a refreshing ocean breeze, the workshop developed in top form. The lectures were outstanding, questions very intriguing, comments and discussion positive and attractive, and lecturers and audience merged together in a unified entity. In view of the above comments, it is rather obvious that this book cannot compete, in principle, with the magic atmosphere breathed at the workshop. For this reason, this book has not been designed to be just the proceedings of the workshop. On the contrary, it is an organized and elaborated version of the workshop's lectures, reflecting a common job between lecturers and editors, in order to present a unified, updated vision of the homeland security facets, accounting for the presence and interaction between threats and countermeasures from one side, and the privacy issue from the other. An overall presentation of this book is now in order.

Chapter 1, "The New Vision of the Homeland Security Scenario" (Shankar Sastry and Giorgio Franceschetti), presents a brief history of the homeland security (HS) scenario, initially limited to defence from terrorists' threats, then extended to early warning of natural disasters, and improvement of the resilience of the underlying cyber world where our commodities usage is implanted, and needs itself security. In the Conclusion section, an even more advanced vision of HS is advocated. It turns out that a global approach to HS needs an enlarged vision, wider perspective, and novel and integrated procedures and techniques to be developed. This implies the necessity of a wide cooperation between the academic and industrial world, together with appropriate laws and policies. In the chapter appendix an ad hoc organization, the Team for Research in Ubiquitous Secure Technologies (TRUST), aimed to the design, building, and operation of trustworthy information systems for critical infrastructures, is presented. A large number of pertinent examples and realizations are discussed: financial infrastructures (including telecommunications and e-commerce), health infrastructures (including personal health records), and physical infrastructures (including the power grid, transportation, oil, gas, and water).

The remaining 10 chapters of the book can be grouped according to the following homogeneous sections.

1. Basic aspects common to the entire HS scenario;
2. Privacy-aware design of sensing systems, aimed at security implementation;
3. Specific interesting and illuminating HS implementations, where the threat is detected and properly counterbalanced.

In regards to point 1, basic aspects common to the entire HS scenario, Chapters 2–4 and 8, are hereafter briefly illustrated.

Chapter 2, "Homeland Security and National Defense in the Twenty-First Century" (Pietro Finocchio and Pompeo Piedimonte) starts from the statement: "There can be no freedom without security" (Wilhelm von Humboldt), and then notices that, after the fall of the Berlin Wall, the global political landscape has evolved towards a new scenario of regional balances of powers. These escalated into a series of local tensions and crises, with high probabilities of asymmetric conflict and attacks, directed against both civilian and military targets. Accordingly, the transformation of the military organizations in the direction of a dual use system to operate for both military and civilian defense is presented, together with the integration of sensors and information sharing inside the communications networks, the so called net-centric doctrine.

Chapter 3, "Homeland Security and Challenges in Information Systems" (Bernhard M. Hämmerli), presents an overview of the information and communication technologies (ICT) role into nearly every component of critical infrastructure, as well as into offices, homes, and even user devices (e.g., cars). ICT has given to our society efficiency and comfort, but has also added complexity and vulnerability. E-banking security and processes are used as an example to illustrate the information security challenges and its associated impact on our society, as addressed in the European Union by the PARSIFAL project. Specific consideration on the balance of security measures and privacy concerns is also given.

Chapter 4, "Analysis of Emerging Phenomena in Large Complex Systems" (Alfonso Farina, Sofia Giompapa, and Antonio Graziano) is concerned about theory, applications, and examples of large and complex organization, encompassing a larger ensemble of actors, hardware, and software components, all cooperating to extend the system's overall capability. This ensemble of multiple simple units interacts nonlinearly among themselves, and exhibits a behavior that is not merely given by adding up the behaviors of the individual components. The chapter attempts to clarify the emergent aspects of this new science of complexity, in relation to the large systems operating in the domains of HS and homeland defense (HD).

Chapter 8, "Privacy Versus Security: A Fight That May Turn into an Alliance" (Adarosa Ruffini and Giorgio Franceschetti), addresses the privacy issue, whose relevance in the HS scenario is continuously increasing. The form and function of privacy, security, and safety, as permitted and prescribed values in the social system of reference, are analyzed, broadening their conceptual perspectives. It is concluded that privacy shares the same need of protection as security and safety, along the same lines of implementation, as anticipated in the title of the chapter. However, it is also noted that this mutual origin requires the establishment of new rules and common agreements in the frame of the international community. Examples of initial movements along this direction are detailed as well.

Now considering point 2, the privacy-aware design of sensing systems, its basic philosophy is detailed in Chapter 9. Then Chapters 5 and 6 analyze the security of relevant critical systems, namely, health information and urban defense, where privacy is one of the key issues.

Chapter 9, "Privacy-Aware Design for Monitoring, Control, and Protection of Critical Infrastructure" (Stephen B. Wicker), begins by illustrating the powerful potential for sensing systems in protecting critical infrastructures. After that it discusses the nature of privacy and the emerging concerns created by advanced sensing systems. A case study focusing on home energy demand and control systems is developed. As shown for the case of power consumption recording, data collection often creates a privacy problem for the individual, and a relations problem for the collecting public authority; a solution in the form of a set of principles for privacy-aware design is introduced.

Chapter 5, "Model-Based Design of Trustworthy Health Information Systems" (Janos Mathe, Jan Werner, and Janos Sztipanovits), starts from the observation that healthcare is rapidly moving towards a delivery system that is based on advanced information technology. This is happening in virtually all aspects of the enterprise, migrating from traditional, paper-based records to health information systems (HIS), a collection of computer-based applications that enables sophisticated services for patients and health care providers. The result is that HIS is forced to manage intricate and dynamic privacy policies, thus requiring extensive ethical, legal, modeling, and policy abstractions research. All these issues are thoroughly examined, with case examples presented. The chapter is enriched by 57 references.

Chapter 6, "Urban Defense Using Mobile Sensor Platforms: Surveillance, Protection, and Privacy" (Mario Gerla, Uichin Lee, Nanxi Chen, and Soon Oh), discusses how video cameras are proliferating at a very fast pace in urban areas all over the world. This chapter presents mobile video cameras mounted on vehicles as an alternative to video cameras installed in the infrastructure (walls, light poles, bridges, and so forth) of the monitored area. In the chapter vehicular platforms and efficient vehicular protocols for surveillance are presented, addressing the unique security and privacy requirements posed by HS applications. A testbed, MobEyes, currently under development at UCLA, is described, which will be used to test and demonstrate the validity of these alternative concepts of surveillance, implemented without compromising the privacy of the citizens.

As anticipated under point 3, specific interesting and illuminating HS implementations are also included in the book. Chapters, 7, 10, and 11, address specific HS and HD implementation, where the threat is detected and properly counterbalanced. Airport security and antipiracy cases are presented, as well as a rather unique security implementation of a major event coupled with a natural disaster happening.

Chapter 7, "Detection and Identification of Dangerous Materials for Airport Security" (Pierfancisco Lombardo, Fabiola Colone, and Roberta Cardinali), introduces the state of the art of the airport security measures, which is very important but very difficult to address. Some of the innovative solutions presently under study, in the frame of the European FP7 Project ATOM, are presented. Appendix 7A, "List of Airport Security Measures Failures with Worst Consequences," summarizes the security failures from 1972 to 2009. Its reading is very impressive, as a testimony of the still large research and experimentation needs for a really efficient airport security implementation.

Chapter 10, "Military Defense, Civil Defense, and Civil Protection Integration in a Multiscenario Crisis Event" (Roberto Mugavero), describes the unique experience carried out during the organization and management of the major event, G8 Summit 2009, which took place in the city L'Aquila (Italy) just 3 months after the earthquake, of a significant size, magnitude (Richter scale 5.8), and effect, which struck the city itself. This experience has led, for the first time, to the development of a methodology and a model of integration in which civilian and military capabilities, aimed at two different aspects of safety, cooperated together. As a matter of fact, civilian and military organizations have joined forces to create a unique and homogeneous framework in which two events, parallel, but, at the same time, converging, take place. It is interesting to recall that the chapter author, in addition to being a university professor, is also volunteer firefighter officer, of Colonel rank, accordingly in the best position to testify about this unique happening.

Chapter 11, "Repel Boarders!" (Patrizia Fabbri and Leonardo Quattrocchi), synthetically presents the piracy scenario, with the aid of interesting and attractive figures, and finally describes an effective and successfully tested system to counter the threat of piracy at sea. This antipiracy system, POMPEIUS, is adequately detailed. This system, at the state of the art in the antipiracy scenario, has been already successfully experimented, and the experiment's results are mentioned.

In conclusion, this book does not reflects only the presentations to the workshop, but an elaborated vision of all the matter, including HS threats, countermeasures, and the privacy issue, as anticipated in its title.

Giorgio Franceschetti
Napoli, Italy

Marina Grossi
Rome, Italy

Editors
May 2011

1

The New Vision of the Homeland Security Scenario

Shankar Sastry and Giorgio Franceschetti

1.1 Unveiling the New Scenario

It is a common assumption that the *homeland security* (HS) concept was born just after the Twin Towers bombing, September 11, 2001, took place (Figure 1.1). This is certainly true at the media level, so that HS and its connected *homeland defense* are automatically related to terrorism activity and terrorists' actions. However, such a definition does not accurately set the HS boundaries, which are much wider and continuously widening. This chapter attempts to present and discuss this issue, leading to a realistic definition of this scenario, and the correct way to handle it. All the chapters of this book are based on the lectures presented, discussed, and elaborated in the workshop [1], and are devoted to examine the different HS facets, thus providing proof of its increasingly spreading within our everyday life.

Actually, there are relatively few books on the homeland security topic, and a large part of them are popular/novel writings. On the technical side, the following titles [2–9] may be quoted. Among these books, [2, 3, 5] are very introductive, and the others (except [9]) are mainly focused, at a target audience of law enforcement and policy makers. Accordingly, to the best of our

Figure 1.1 Photograph of the September 11, 2001, Twin Towers bombing. (© Associated Press/LaPresse).

knowledge, only [9] is devoted to a high-level audience. This book, and this chapter, is dedicated to this same audience, with the different perspective of an improved definition of the HS scenario.

To start with, it is convenient to recall a few terrorist attacks, in order to present techniques, motivations, and targets. These case histories are presented hereafter.

1.1.1 Tokyo Chemical Sarin Attack

The Tokyo Chemical Sarin Attack happened on the Tokyo subway on March 20, 1995. The attack occurred at about 8:00 a.m., when traffic was busiest. Five two-person teams entered the Tokyo (Japan) subway, each team carrying a sealed camouflaged package, that was deployed at different places in the subway system, punctured with sharpened umbrellas, allowing Sarin liquid (a chemical nerve agent) to ooze into the subway cars and gradually vaporize. The attack caused 12 deaths and contaminated 3,500 to 5,500 people, with 1,000 people requiring hospitalization. During planning and preparation for the attack, about 1 year before the city of Matsumoto (Japan) suffered a similar Sarin attack in one of its residential neighborhoods. The terrorists previously developed their capability to produce Sarin, by investing several million dollars and hiring scientists and skilled workers to start a facility (Satyan) in Japan.

1.1.2 Murrach Federal Building Attack

The Murrach Federal Building Attack happened in Oklahoma City (United States) on April 19, 1995; the truck bombing occurred at 9:02 a.m. The truck was positioned at the delivery access point of the Murrach Building site, used by several agencies of the United States (Department of the Army, of Defense, of Agriculture, and so forth) during the past years. The driver stepped out the truck, and disappeared; a few minutes later the bomb went off. The building collapsed, killing 167 people (including 19 children) and injuring 853. The bomb was constructed by the terrorist (and his team) by using 5,000 pounds of ammonium nitrate fertilizer mixed with 1,200 pounds of liquid nitromethane, and other components available in department stores; the estimated cost was $5,000.

1.1.3 Beslan

The Beslan hostage crisis and mass murder happened in North Ossesia (Caucasus region) on September 1, 2004. A group of about 30 terrorists, camouflaged and heavily armed, drove into a school courtyard of Beslan around 9:00 a.m., jumped from vehicles, and occupied a school full of students and their parents. They were barking orders to the hostages, killing those that did not accomplish their requests. Local police and other security forces gradually cordoned the school around 10:30 a.m., and negotiations started and continued for 3 days; the issue was the Chechnya independence, a region close to North Ossesia. The third day the gymnasium of the school erupted in a large explosion, and security forces entered the school with their tanks, fighting with the terrorists, room by room, in the school. The terrorists' casualties amounted to 20, while over 300 men, women, and children were killed in this mass murder.

1.1.4 London Bombing

The London Bombing occurred in the London transportation system on July 7, 2005. At about 8:50 a.m., within seconds of each other, three suicide terrorists detonated bombs on there subway trains, as the trains departed into tunnels of the London subway system (the Tube). A short time later, at 9:47 a.m., a fourth suicide bomber detonated another bomb on a double-decker bus in the snarled traffic of a busy London street. The four suicide bombing resulted in 56 deaths, and about 350 people needing hospital treatment.

1.1.5 Penetration of the U.S. Electric Grid

The U.S. electric grid was penetrated on April 8, 2009. The U.S. electrical grid comprises three separate electric networks, covering the East, the West, and Texas. The growing reliance of utilities on Internet-based communication has increased the vulnerability of control system to hackers and possibly spies. A heavy intrusion was detected on April 8, 2009, but the intruder, or intruders, were not detected; were they Chinese spies or Russian spies? (Both heavily denied this.) Were they other spies? Most importantly, were software programs left behind, which could be used to disrupt the U.S. electric system?

Examination of even only these few cases identifies some key parameters entering into an attack: use of different *weapons*, as chemical agents; homemade exploding devices, like professional ammunitions up to small tanks; intended *targets*, like a mass of people or electricity distribution; and specific *motivations*, like scaring people, offense to national institutions, or recurrences. A more ample, but certainly not exhaustive, information of threat parameters is graphically depicted in Figure 1.2; the conclusion is that the variety of attacks is very large. Classification of the different types of threats would be very desirable, for

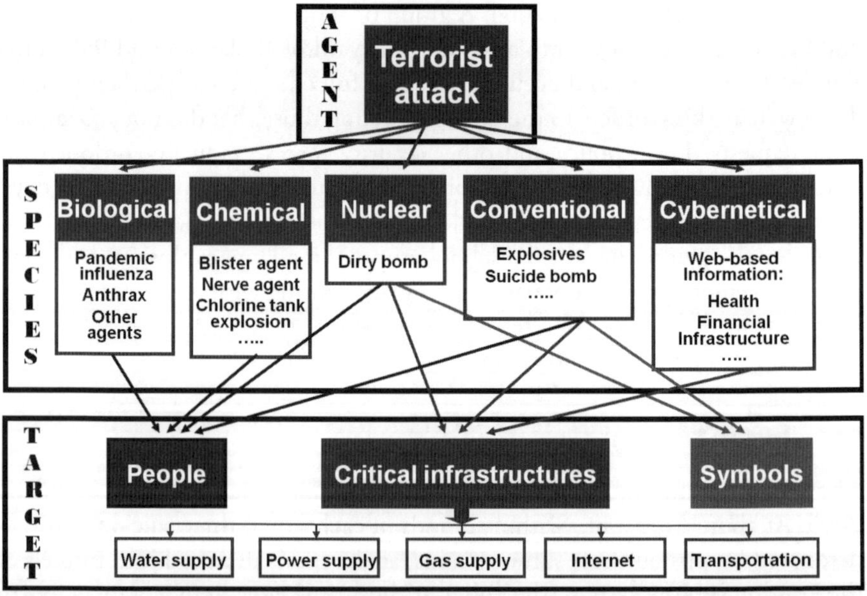

Figure 1.2 Graphical representation of a terrorist attack. Several types of adopted weapons and targets are indicated: the former are biological, nuclear, cybernetic, conventional, and chemical as weapons; the latter are people, critical infrastructures, and symbols as targets.

the generation of statistics, and very useful for the design of appropriate countermeasures. But this is no easy matter. Several attempts, sponsored by U.S. Department of Defense, and National Counter Terrorism Center (established by #13354 U.S. presidential order, 2004), did not lead to totally reliable and comparable results.

After the September 11, 2001, World Trade Center bombing, the HS activity significantly increased, along a wider perspective: development of procedures and techniques for early warning of the threat, in addition to effective countermeasures. It was soon realized that the former activities were certainly also effective to mitigate danger and causalities in the event of natural disasters (earthquake, volcanic eruption, landslide, and so forth). Accordingly, the previous attempts in this area, namely geophysical early warning [10], were absorbed into the homeland security discipline. This extension of the HS scenario is not the last one; a recent case is very indicative.

1.1.6 Vandals Cut Phones and Internet Cables

In April 2009, vandals cut the phone and Internet cables in Morgan Hill, California. Up to four fiber-optic lines were intentionally cut around 2:00 a.m. in south San Jose, resulting in serious service problems across Santa Clara, San Benito, and Santa Cruz. The result was a total mess: credit cards were not accepted, automatic cash machines were not working, ambulances could not be requested in emergency situations, and so forth. The police chief made a synthetic statement, saying that: "People won't know what to do without their phones." This situation lasted till 6:00 p.m. of the same day.

The idea that some excited kids, coming out of a club, may easily impair the functioning of our life system and our commodities usage is quite disturbing. The *underlying cyber world*, where our commodities usage is implanted, is very delicate, probably unstable, and needs security. And a very recent occurrence adds additional worries to this security issue.

1.1.7 Stuxnet Computer Worm

The Stuxnet computer worm was discovered by VirusBlokAda, a security firm in Belarus, in July 2010. Stuxnet is an Internet worm that infects Windows computers. It primarily spreads via USB sticks, which allow it to get into computers and networks not normally connected to the Internet. Once inside a network, it uses a variety of mechanisms to propagate to other machines within that network, and gain privilege once it has infected those machines. While it is not the first time that hackers have targeted industrial systems, it is the first discovered worm that spies on and reprograms industrial systems, and the first to include a programmable logic controller (PLC) rootkit; these are small embedded industrial control systems that run all sorts of automated processes: on factory floors, in chemical plants, in oil refineries, at pipelines—and, yes, in nuclear power plants. Stuxnet includes the capability to reprogram the PLCs and hide its changes.

A lot of press and media speculation has developed over the Stuxnet story. Some theories include: this worm might have damaged Iran's nuclear facilities in Natanz, and eventually delayed the start up of Iran's Bushehr Nuclear Power Plant; the attack was lead by Israel; the virus infection spread over China; a wide numbers of computers have been infected all over the world, including United States and Europe; and so forth. Although these are speculations of questionable reliability, one point is scary: prevention requires a multilayered approach, often referred to as *defense-in-depth.* The layers include policies and procedures, awareness and training, network segmentation, access control measures, physical security measures, system hardening (e.g., patch management), and system monitoring, antivirus sophisticated procedures. In the absence of appropriate hardening of our systems, the society that we constructed over the centuries is in serious danger; and even false news about security problems may generate, via omnipresent media, a wave of terror, with unpredictable consequences.

A preliminary conclusion is in order. In ancient times, our life was essentially confined in the $\alpha\gamma o\rho\alpha$ (agorà), an ancient Greek word indicative of the *square* of the village; the social environment was very limited, but very stable. The situation is now just the opposite. We live in a *global village,* whose roots and foundations are delicate and possibly unstable. It in necessary that these roots are made *resilient,* irrespective of the existence of terrorists and/or natural disasters. The updated meaning of the acronym *HS* is now to be read as *homeland stability.* Novel procedures and techniques for its implementation must be constructed. In Section 1.3, an even more advanced vision of HS is advocated.

1.2 The New Rules and Operational Avenues

At the end of the down-to-earth considerations of Section 1.1, we have realized that the operational roots of our society and way of living rely on critical systems, as (but not limited to) electricity, transportation, banking, telecom, and health care. These system are not isolated, but rather interconnected, since a failure of one systems may affect, and usually affects some, if not all the other ones. Operation and connection of the systems, or rather of this composite *system of systems,* is made via wired and wireless transmission and sharing of information, along appropriate software in charge of computers. In order to render our life safe and productive, we should protect the critical systems upon which our economy and society depend, and assure the *cyber security* against its intrinsic and/or intentional human misuse. In particular, we should prevent intrusions, detect intrusions (if happened), limit damage, and operate through attacks. This policy may be summarized in a single statement: critical systems must be designed to be *trustworthy*. These are systems that do what is required despite environmental disruption, human user and operator errors, and attacks by hostile parties. In a summary, they should provide:

- Strong enough *barriers to penetration*;
- Accurate *intrusion detection*;
- Ability to *fuse incident reports* across a global area and *deduce possible* plans for warning and to guide interventions;
- Tolerate attacks and *keep on ticking*.

These requirements need a full technology knowledge and operational ability, to be essentially classified into five areas:

- Information assurance and survivability;
- Security with privacy;
- Secure network-embedded systems or cyber physical systems;
- Validated modeling, simulation, and visualization of critical infrastructures and their interdependencies;
- Public-private partnerships for technology transition.

We emphasize the security-privacy merging, because technology solutions that feature strong privacy safeguards actually help strong security. Modeling is also of paramount importance: it is the first necessary step to realize a simulator, a fundamental device not only for the system design, but also its real-time adjustment to sudden changes of the environment and threat. Last, but certainly not least, the public-private partnership is a key element in the attempt to protect the critical systems upon which our economy and society is depending: this issue is not steered by business or market; it is what our overall society needs.

Above-mentioned requests are reasonable and convincing. But the obvious question is: how can they be implemented? It is obvious that new patterns must be explored. In the following we introduce an organization, Team for Research in Ubiquitous Security Technology (TRUST),[1] which was created in

1. Appendix 1A details TRUST recent activities and accomplishments.

response to these challenges to focus on the development of fundamental cyber security science and technology.

TRUST is a consortium of six universities (Berkeley, Carnegie Mellon, Cornell, San Jose State, Stanford, and Vanderbilt), which provides a framework within which the team collaborates to solve problems far too large and complex for any individual group to tackle. TRUST has already achieved success along several fronts and is addressing fundamental scientific and technological problems, advancing the state of the art in a number of areas:

- Security and privacy issues associated with the rapidly increasing use of electronic media for the archival and access of patient medical records;
- Web authentication, end-user privacy, next-generation browser security, malware detection, and improved system forensic techniques to combat online attacks;
- Application defenses for network-level intrusions and attacks, including compromised and malfunctioning legacy applications, viruses, worms, and spyware;
- Laws and policies that combine with market incentives align the design, deployment, and configuration of systems with privacy, security, and trustworthiness goals;
- Secure embedded sensor networks for large-scale applications critical to the nation's economy, energy, security, and health;
- Techniques that ensure trustworthy computing by securing hardware, improving software robustness, and increasing the survivability of critical systems.

In its first 5 years of existence, 2006–2010, TRUST used a rigorous strategic planning process for identifying the key challenge problems. Its experience has been consolidated by proceeding inductively from studying three grand challenge applications: *financial infrastructures* (including telecommunications and e-commerce), *health infrastructures* (including personal health records), and *physical infrastructures* (including the power grid, transportation, oil, gas, and water). Students at the undergraduate and graduate levels have actively participated in the research in the labs, have been exposed to novel teaching material, and have integrated the best new research into classical pedagogical procedures. Tools, techniques, and other research results have been properly disseminated to ensure that achieved accomplishments transitioned from the laboratory to the end users.

It is evident that only an integrated structure as TRUST can successfully operate on challenge problems, comprising network-embedded systems, end-user security, trustworthiness in electronic medical records, trusted systems and software, networking, and privacy-security combined issues. In addition, only a multifacet structure, as TRUST, may translate the resulted substantive multidisciplinary progress to vendors, infrastructure service providers, end-users (individuals), government organizations, international partners, and others. This successful experiment may be a guide for similar ones. For this reason, at the end of the chapter an Appendix is attached, where the TRUST experiment is described, trustworthy accomplishments are detailed, and a large reference list (more than 100 titles) is cited.

We believe that the right way to proceed within this extended HS scenario is by implementation and usage of highly integrated structures, TRUST being a valid prototype, which present valid realizations as well as fundamental future research areas. As in the TRUST case, the competences should be very wide, spanning from the technical to the legal side; and the connection and inclusion of public partnership, essentially in charge of the final decisions from the political side, is another significant feature.

1.3 The Last Step

In the two previous sections the historical trajectory of HS, from the defense against terrorists' attacks to the mission of stabilizing the roots of the critical infrastructures at the basis of our society, has been presented. And again a question is raised—is this the end of the HS trajectory? If the meaning of the word *homeland* is the conventional one (i.e., national territory, people, goods, and so forth) the trajectory reached its dead end. But if we enlarge the meaning of this word, the journey may be further in progress.

There is a nice word in Italian, *patria*, originating from the Latin word "pater" (i.e., "father") that is indicative of the country, but implies not only the homeland commodities, but also its culture, traditions, and all the ideal accomplishments that flourished in that land over the centuries. Homeland security should also take care of preserving these values, which are invaluable inheritance of the country. These values must be saved against any brainwash threat to the mental freedom of the people. The "enfants de *la Patrie*" ("the sons of the Nation," reference is made to the French national song) were fighting, during the French Revolution, against the French nobility at that time, no more representative of the national spirit. And we should also be sure that the identity of our nation is, and will be preserved forever, avoiding any *homogenization process* aimed to turn off diversities in favor of a uniform gray landscape all over the world. The right weapon for this accomplishment is the culture, which should be increased all over the population: continuing education, with the aid of competent educators, at all levels and all ages must be assured and implemented. This is a dream, but perhaps is going to become a necessity over the years. Then, we will reach the final reading of the acronym *HS*, which is *homeland serenity*.

And the story of the 3S—*security, stability, serenity*—will be finally over.

References

[1] "Homeland Security Facets: Threats, Countermeasures, and the Privacy Issue," workshop held at Baia (NA), Italy, September 21–25, 2009.

[2] Ranum, M., *The Myth of Homeland Security*, New York: John Wiley, 2003.

[3] Sauter, M., and J. Carafano, *Homeland Security: A Complete Guide to Understanding, Preventing, and Surviving Terrorism*, New York: McGraw-Hill, 2005.

[4] Kamien, D., *The McGraw-Hill Homeland Security Handbook: The Definitive Guide for Law Enforcement, EMT, and All Other Security Professionals*, New York: McGraw-Hill, 2005.

[5] White, J. R., *Terrorism and Homeland Security: An Introduction,* Boston, MA: Wadsworth Publishing, 2008.

[6] Kettl, D., *System Under Stress: Homeland Security and American Politics,* Washington DC: CQ Press, 2007.

[7] Haddow, G., D. P. Coppola, and S. Yeletaysi, *Introduction to Homeland Security: Principles of All-Hazards Response,* 3rd ed., Boston, MA: Butterworth-Heinemann Ltd., 2008.

[8] Chertoff, M., and L. H. Hamilton, *Homeland Security: Assessing the First Five Years*, Philadelphia, PA: Penn Press, 2009.

[9] Franceschetti, G., and M. Grossi, (eds.), *Homeland Security Technology Challenges: From Sensing and Encrypting to Mining and Modeling*, Norwood, MA: Artech House, 2008.

[10] Gasparini, P., G. Manfredi, and J. Zschau, (eds.), *Earthquake Early Warning Systems,* New York: Springer-Verlag, 2007.

Appendix 1A
Team for Research in Ubiquitous Secure Technologies (TRUST)

S. Sastry, L. Rohrbough, K. Gates, R. Bajcsy (University of California, Berkeley), S. Meldal (San Jose State University), J. Mitchell (Stanford University), D. Mulligan (University of California, Berkeley) A. Perrig (Carnegie Mellon University), F. Schneider (Cornell University), J. Sztipanovits (Vanderbilt University), D. Tygar (University of California, Berkeley), and S. Wicker (Cornell University)

1A.1 Introduction

Keeping information and computer systems secure is one of the country's most formidable challenges. Daily headlines chronicle fraud, viruses, privacy abuses, network outages, and security vulnerabilities across the spectrum of critical systems that support the nation's financial, energy, healthcare, and transportation infrastructure. Many studies and most recently the 2009–2010 Comprehensive National Cyber-Security Initiative from the White House identify security and robustness of the critical infrastructure as one of the most urgent challenges facing the nation. We have seen a rapid increase in computer security attacks at all levels of our critical infrastructures, as more individuals use computer networks, and as the motivations and means by well funded professional cyber hackers lead to a significant increase of sophisticated attacks. Our laws and regulations remain entrenched in traditional crime-fighting approaches to cyber security, failing to reflect the complicated and sometimes indirect ways in which security and privacy failures arise and the range of interventions—aimed at improving our capacity to prevent, detect, deter, contain, and recover from security failures—that are essential to the evolution of more trustworthy systems [1–101].

The Team for Research in Ubiquitous Security Technology (TRUST—see http://www.truststc.org) was created in response to these challenges, to focus on the development of fundamental cyber security science and technology that will radically transform the ability of organizations to design, build, and operate trustworthy information systems for the nation's critical infrastructures. TRUST was formed to bring together a visionary cyber security and infrastructure protection systems research team unparalleled in its strength and accomplishments. Drawing on the expertise at Berkeley, Carnegie Mellon, Cornell, San Jose State University, Stanford, and Vanderbilt, TRUST provides a framework within which the team collaborates to solve problems far too large and complex for any individual group to tackle. TRUST works to build a science base for trustworthy systems as well as addressing technical, operational, legal, policy, and economic issues affecting security, privacy, and data protection. TRUST has already achieved success along several fronts and is addressing fundamental scientific and technological problems and advancing the state of the art in a number of areas: (1) security and privacy issues associated with the rapidly increasing use of electronic media for the archival and access of patient medical records; (2) Web authentication, end-user privacy, next-generation browser security, malware detection, and improved system forensic techniques to combat online attacks; (3) application defenses for network-level intrusions and attacks, including compromised and malfunctioning legacy applications, viruses, worms, and spyware; (4) laws and policies that combine with market incentives to align the design, deployment and configuration of systems with privacy, security and trustworthiness goals; (5) secure embedded sensor networks for large-scale applications critical to the nation's economy, energy, security, and health; and (6) techniques that ensure trustworthy computing by securing hardware, improving software robustness, and increasing the survivability of critical systems.

A criticism frequently made of cyber security research is that it is too reactive to problems in the headlines. A similar criticism can be leveled at laws aimed at improving security and privacy. The result is an *arms race* between progressively more sophisticated attacks followed by more innovative defenses and an ongoing race to apprehend the wrongdoers. TRUST has embarked on a different course by attempting to develop a new science and technology base for trusted systems, and change this reactive mindset to a proactive one. We expect to develop this science base by proceeding inductively from studying three grand challenge applications: *financial infrastructures* (including telecommunications and e-commerce), *health infrastructures* (including personal health records), and *physical infrastructures* (including the power grid, transportation, oil, gas, and water). These application areas were selected because they represent the most important trustworthiness problems confronting U.S. society, and more generally industrialized societies worldwide today; each of them emphasizes fundamentally different kinds of trustworthiness issues, each allows us to explore legal and policy interventions that might improve trustworthiness, and each is poised to make increasing use of networked computing, where trustworthiness issues could be a substantial impediment to success, if not adequately addressed.

1A.2 Research Overview

In the first 5 years of TRUST, we have worked on challenge problems in network-embedded systems, end-user security, trustworthiness in electronic medical records, trusted systems and software, networking, privacy, and security. These resulted in substantive multidisciplinary progress and translation of our results to vendors, infrastructure service providers, end users (individuals), government organizations, international partners, and others. Building on this and going forward in the next five years, we expect the legacy of TRUST to be the development of a new science of cyber security by abstracting inductively fundamental game-changing solutions for three grand challenges: trustworthiness of three key critical infrastructures: *trustworthy financial infrastructures, trustworthy health infrastructures,* and *trustworthy physical infrastructures.* The choice of grand challenge areas emerged from their national importance and early work of the Center. For example, the choice of trustworthy physical infrastructures is motivated by our early successes in secure wireless network embedded systems, and the security and privacy of wireless sensor networks. We believe that these wireless sensor networks will be the scaffolding for the next generation information infrastructure for physical infrastructures such as in electric power, transportation, water, oil, gas, and so forth [so-called supervisory control and data acquisition (SCADA) systems]. For example, the next generation of SCADA will be built on top of the secure wireless embedded sensor network work being done in the TRUST Center. The Center research on end-user security, privacy preserving password enhancements and audit technology for distributed systems will be a key to Trustworthy Financial Infrastructures and ensure the trustworthiness of e-commerce, telecommunications, and financial infrastructures. Finally, current TRUST research on security and privacy of (real-time) personal health records, electronic medical records with special regard to HIPAA compliance, privacy policy audits will be the foundations for the work in trusted health infrastructures. The grand challenge applications were specifically chosen to help develop a fundamental science and technology base for trustworthy systems and additionally to show how the core technologies developed in the Center can be integrated into *systems-of-systems solutions,* which will transcend the endless *arms race* mode, which is the current defense of our critical infrastructures. Indeed the three infrastructures chosen represent by most studies the major national challenges. Thus, we are simultaneously trying to develop the game-changing cyber security technology tools and technologies from fundamentals, while integrating them into solutions for some of the pressing problems confronting us today.

This appendix provides an overview of the three TRUST grand challenge research areas. Specifically, it presents the outcomes of three detailed studies that we have performed over the last years in the style of a Defense Advanced Research Projects Agency (DARPA) Information Science and Technology (ISAT) inspired study format. Our three studies are aimed at answering the so-called *Heilmeier questions,* such as: What are we trying to do? If we succeed, who will care? Why do we think we have the tools and experience to be successful? What are the midterm and final challenges (in terms of translating the research outcomes)? We describe hereafter the outcome of these three studies.

1A.2.1 Financial Infrastructures

This thrust of TRUST aims to develop the central science and engineering principles needed to support the long-term future security, reliability, and ubiquitous usage of the financial infrastructure. This comprises financial service enterprises, online retail businesses, and their customers, linked together in a trustworthy environment supporting commercial transactions. Its four key components are:

- *Customers:* Individuals interact on personal computers with financial service providers through e-mail and the Web. The individuals are usually not technology experts, nor are they able to afford professional systems administrators, yet they need to be assured of reliable interaction.
- *Service providers:* Financial service providers and online retailers interact with customers through e-mail, operate Web servers, and carry out back-office operations subject to rigorous security and performance requirements. They also have complex partnering agreements, and rely on their brand image and reputation for competitive advantage.
- *Interconnection:* Financial infrastructure customers rely on open networking standards, browser architecture, and Web application development practices. Providers may also communicate through private networks, leverage federated identity management solutions, and outsource functions to other providers through complex networking practices.
- *Policy:* Financial services and online retail enterprises are subject to complex and overlapping regulations, evolving levels of customer awareness, and differing sensitivities. The mix of policies, necessary to fuel security in an increasingly decentralized system in which consumers with relatively limited technical expertise and limited desire to manage security and privacy play a central role, is essential to explore during considerations of technology.

The massive volume of financial transactions makes it critical for us to have a trustworthy infrastructure for keeping consumer confidence in financial institutions. Further, the factors listed above make this problem much more challenging than traditional distributed systems. A common taxonomy of security enforcement mechanisms speaks in terms of a gold standard of trustworthiness, because the first letters of each element are the atomic symbol for gold (Au): authentication, authorization, and audit. Inadequate means of authentication is responsible for much of the identity fraud and theft seen today, including phishing attacks. Not only must the financial institution authenticate the customer, and have incentives to use appropriate mechanisms for doing so, but the customer must also know how to authenticate the financial institution to avoid phishing, pharming, and other criminal attacks. This is a rich problem space, where TRUST researchers have already made very substantive progress accompanied by the transition of important new products such as Pwd Hash and Dynamic Security Skins [1, 2], but much more remains to be done. As enterprises develop richer Web-based applications, we see a vast landscape of authentication, authorization, and session integrity challenges, stemming from differences between browser platforms, server platforms, and software development tools and technologies.

1A.2.2 Health Infrastructures

Healthcare has been characterized as a *trillion-dollar cottage industry* that is still dependent upon paper records and fragmented, error-prone approaches to service delivery. According to a landmark National Academies Study [3], over 98,000 Americans have died, and more than 1 million patients have been injured due to process and system failures in health care. The Picker Institute survey [4] pointed out that inefficiencies have resulted in a double-digit growth in the cost of health care, and a large increase in the uninsured population in the United States. However, there is an emerging vision for the future of healthcare [5] that can bring dramatic change in the quality and cost of health care. This vision is not only a matter of correct policy or business decisions. According to TRUST researcher Dan Masys, there are several forces at work that push the healthcare industry towards radical change [5], including: the dramatic increase in the amount of information required for making health decisions, the rapidly growing use of the Internet worldwide, genome research that opens up the opportunity to provide personalized healthcare, and medical errors caused by failures in information management. It has been increasingly recognized that IT can indeed make a huge difference by enabling the creation of disruptive technologies that can change health care [6]. Personal health records (PHR) are predicted to be a game changer by creating informed and engaged patients. The future importance of input from policy makers to PHRs has been recognized by the large IT companies with their service offerings, such as Microsoft's HealthVault and Google Health.

A revolution has started up in real-time medical decision support systems as well. Increasing numbers of early commercial instances of decision support systems are being developed with new capabilities. For example, recent advances in wireless sensor networks have made it possible to deploy wearable sensors on the bodies of patients in an at-home setting, allowing continuous monitoring of physiological signals (such as ECG, blood oxygen levels), and other health related information (such as physical activity levels). This technology offers a unique opportunity for shifting healthcare outside a traditional clinical setting to a patient/home-centered setting. Progress in these new, transformational applications of IT in health care is predicated on the availability of technical solutions for some extremely hard problems: implementation of privacy requirements in health information systems (HIS), and the availability of design technology that can guarantee safe operation in the rapidly changing health care environment.

TRUST researchers have started addressing some of the fundamental issues driven by the major technical goal of TRUST: *to design trusted systems, which are composable from component technologies.* A primary concern in HIS design is that privacy and security requirements are frequently expressed in vague, as well as contradictory, complex laws and regulations. Development of formal policy modeling languages and analysis methods is one of the early successes for TRUST. Engineering of complex software systems that are functionally complete, able to adapt to the changing healthcare environment, and can adopt to changing security and privacy related laws and regulations is hard, if not impossible to solve using conventional software and systems design technology. To address this problem, TRUST researchers introduced model-based methods to offer a revolutionary way to formally and explicitly integrate privacy and security goals into HIS architectures [7–12]. While we have achieved progress both in terms of problem understanding and developing new foundations, we placed strong emphasis on experimental work. Taking advantage of our partnership with the Vanderbilt Medical Center, we have developed a test bed for model integrated clinical information systems (MICIS), and for home-based health monitoring, which integrates our research results in a tangible platform and is used by the medical community for testing and validation [13].

1A.2.3 Physical Infrastructures

The focus of this area is next generation secure supervisory control and data acquisition (SCADA) and other networked embedded systems that control critical physical infrastructures (e.g., power grid, natural gas distribution, automated railroad control, water, transportation) and more futuristic infrastructures such as "smart (zero energy) buildings" and "smart structures" (e.g., active bridges whose structural integrity depends on dynamic control or actuators to replace static brute-force physical structure). Unlike the other grand challenge areas, there is virtually no legacy system problem to overcome in physical infrastructures using new secure SCADA systems and built on top of the emerging new technology of wireless networked embedded systems. However, there are substantive issues of ownership of the control of the physical infrastructure (whether it is individuals inside their homes or the utility provider in the case of the grid).

Security requirements are traditionally enumerated in terms of confidentiality, integrity, and availability. In this challenge area, confidentiality is not nearly as important as the other two dimensions. Moreover, availability is often too weak—real-time constraints must be satisfied, which changes the approach for defending against denial of service attacks. Another unique challenge in this area is dealing with the potential of physical compromise of unattended nodes deployed in the field—reliable operation of critical infrastructures needs to be ensured even in a case where the adversary controls a subset of the devices. Privacy issues also arise in this area, but in subtle ways. Most people do not think about what could be inferred from their uses of infrastructure since information is revealed in indirect ways (for example, increased power draw implies somebody is at home). Moreover, when distributed networks of sensors are widely deployed, then opportunities for privacy abuse arise—again, through abuse of information that is being collected (presumably) for other reasons. The more futuristic infrastructures such as smart buildings and structures portend immense data collection in places routinely occupied by individuals. There is a strong opportunity to ameliorate privacy concerns by considering them early on in the technical work and in the accompanying policy and consumer protection research that will support this future.

1A.3 TRUST Research in Secure Financial Infrastructures

The complexity of scientific and engineering challenges facing the financial infrastructure stems from several characteristics of the current environment. Foremost among them is that *attacks against or within the financial infrastructure are prevalent and lucrative.* The FBI estimates that computer crime costs industry $400 billion per year, with estimates of $50 billion for ID theft [14]. Another important characteristic that distinguishes this area from other TRUST grand challenge areas is that *financial systems are not under control of one organization*: Web browsers that execute critical parts of current Web applications are separately administered by nonexperts. In addition, the intra-enterprise financial infrastructure is highly networked. In contrast to traditional computer systems, financial systems *critically involve computers and people.* While authentication of computer systems to each other has been widely studied, Web sites want to authenticate a person, not a machine. In addition, the importance of the human in the loop leads to significant legal, social, policy, and human factors issues. Finally, the financial infrastructure operates in the face of r*apid technological evolution.* Server development frameworks are similarly rapidly evolving, and the rise of ubiquitous handheld platforms provides a means for development and deployment of new technologies that will replace old ones rapidly. In order to better understand the needs of key participants in the financial infrastructure, TRUST interviewed 20 chief information security officers (CISOs), an industry group called The Financial Services Sector Coordinating Council for Critical Infrastructure Protection and Homeland Security, and other sources.

Although different companies see different threats to their internal operations, consensus emerged on some of the most pressing current problems:

- *Authentication:* Financial infrastructure enterprises face challenges in reliably authenticating clients (customers) to site and sites (enterprises) to clients, for both e-mail and Web.
- *Malware:* Enterprises face sophisticated direct malware-based attacks to their information systems, and indirect attacks through malware on their customer's computers.
- *Internal operations:* Enterprises face policy, compliance, and risk management challenges as well as continuing exposure to insider threats.

In addition to these immediate needs, which we believe motivate fundamental scientific study, discussions with financial infrastructure enterprises have led us to identify several broader needs:

- Fundamentally stronger approaches to trustworthy systems to reduce their vulnerability;
- New security architectures for end-user machines not administered by enterprises;
- Greater sophistication in detecting and defending against crime-ware, phishing, malware, account takeovers, code vulnerabilities, authentication, and authorization;
- Trust relationships that are matched with appropriate access control and monitoring mechanisms to: (1) combat insider threats; (2) ensure compliance with government regulatory and internal corporate policy; and (3) allow data mining consistent with consumer protection, and policy norms.

Informed by our discussion and fact-finding efforts, TRUST has formulated a scientific vision and response that will address the long-term fundamental needs described in the following subsections.

1A.3.1 Design of Core Systems Support for Financial Enterprises

The middleware and distributed operating system infrastructure, supporting next-generation financial trading and medical applications, must handle millions of events per second, minimize event delivery latencies and jitter, and maintain system stability and reliability. To improve scalability and support new operational requirements, many applications rely on event-driven middleware platforms that support advanced qualities-of-service (QoS) properties, such as event traffic prioritization, event traffic shaping, hardware and software event filtering, content-aware event delivery, and persistence. In one sample effort in this direction, we propose to demonstrate how to dramatically improve the scalability of real-time event notification systems by integrating Ricochet, a scalable reliable multicast that was recently developed by Ken Birman at Cornell, with the Data Distribution Service (DDS), which is a QoS-enabled event-based middleware platform that enables applications to communicate by publishing information they have, and subscribing to information they need in a timely manner.

1A.3.2 Applicable Network Security for Financial Infrastructure

It is becoming increasingly difficult to ensure security and availability of network operations in today's highly interconnected environments. In one sample proposed effort, we will build on two complementary approaches we have developed for evaluating protocol design. One is a model-checking method that finds errors in protocols. The other is a proof method that can be used to prove security properties of network protocols. We propose to establish mathematical connections between logical and computational methods for protocol analysis, bridging the gap between formal methods and cryptography. In previous work on model checking and Protocol Composition Logic (PCL), a method for proving security properties of network protocols, TRUST researchers identified serious security vulnerabilities in the IEEE 802.11i wireless security standard and the IETF GDOI standard and modern cross-domain Kerberos variants using public-key infrastructure or Diffie-Hellman key exchange. The suggested fixes have been adopted by the respective standards bodies (see http://crypto.stanford.edu/protocols/).

1A.3.3 Tailsearch in Secure Web Security

Cyber criminals are relentlessly inventing increasingly sophisticated Internet attacks [15]. One countermeasure is to improve the internal structure and security policies of modern browsers and leveraging new OS primitives, trusted computing hardware support, and virtual machine monitors. A second approach is to enhance the Web experience with extended tools that provide much higher degrees of protection, in particular, by deferring a number of phishing-based attacks [16–19]. A sample effort will explore Web attacks, develop defenses, contribute defenses to open-source browsers, and develop formal models of the browser security architecture and code executed in the browser that will help find attacks or prove their absence. We believe that a model exposing the policy aspects of browser design, extended to cover aspects of the Web, will allow a larger scientific community to contribute to browser and Web security. Understanding how users are subject to Web attacks involves studying attacks on real users. We have done this and deeply analyzed phishing attacks [18–20], using the knowledge we have learned to develop better browsing techniques [1, 2, 15–17].

1A.3.4 Code Analysis, Monitoring, and Malware Detection

In the area of practical techniques that deeply check software systems for errors, current TRUST research uses a form of symbolic execution to automatically generate tests that (ideally) cover all interesting program paths. We are building a system, klee, based on this approach. In a more specialized direction, we believe that malicious code analysis is extremely important for financial infrastructures. Attackers continuously employ creatively crafted malicious code to attack the financial infrastructures for illegal financial gains. Thus, the ability to automatically dissect a malicious program and extract information from it is an important cornerstone for forensic analysis and defense in financial infrastructures. We plan to leverage our previous work to design and develop the next generation infrastructure for in-depth malicious code analysis and defense, which will be particularly beneficial to the financial infrastructure. We have shown how to attack most learning algorithms [21–29], currently used for malware detection by malicious training of secure learning systems [21].

1A.3.5 Public Policy, Decision Making, and Risk Management

One important set of policy questions are those surrounding the use of personal information by financial enterprises. There is a real question about protecting user privacy [30–35]. On the supply side of personal information, Internet users provide personal information to a wide variety of Web sites for a multitude of purposes: to purchase goods, manage finances, search for jobs, date and meet new friends, and to socialize, to name a few. Many users believe the information is theirs, and thus experience dissonance when personal information has been shared or sold to third parties, when Web sites fail to delete personal information even after the user assumes it should have been, or when hackers or criminals appropriate their information for illegal deeds.

In another sample direction, we proposed to develop connections between computer security and risk management. One illustrative example is the use of SSH for connecting clients to servers. As usually deployed, SSH is vulnerable to a man-in-the-middle attack the first time that a client connects to a server; after the first connection, the client stores cryptographic keys that support authentication on subsequent sessions. A risk-based analysis of the decision to use SSH involves analyzing network characteristics in order to calculate the probability of a successful attack, combine with a decision analysis involving the risk tolerance of the enterprise. We will then propose and analyze new game structures that can better capture the incentive dynamics of new classes of network security threats (e.g., botnets and DDoS attacks).

1A.4 TRUST Research in Secure Health Infrastructures

Over the past decade, many healthcare organizations have started embracing information technology. Since 2002 more than 90% of the approximately 5,000 member institutions of the American Hospital Association have reported having Web sites, with most having descriptive information about their facilities and services. A relatively small but growing fraction of healthcare organizations have created *patient portals* that provide secure, personalized customer services via the Web. For example, Vanderbilt University's patient portal is one of the more advanced healthcare sites, providing a growing set of individualized services to more than 35,000 enrolled patients. In Europe several national initiatives have been started to provide platforms for shared electronic health data records. For example, health@net is an Austrian initiative to develop concepts and an implementation of distributed crossinstitutional health data records. The platform is targeted to support cooperation and information exchange between stakeholders in the healthcare domain like hospitals, family physicians, and pharmacies. These developments and experiences have resulted in the establishment of national goals in HIS [5] that include: archiving and accessing personal medical records, evidence-based personalized healthcare, and home-based healthcare delivery. The HIS project of TRUST has the following three objectives:

1. *Privacy modeling and analysis:* Development of logical foundations and practical methods for expressing and enforcing privacy policy and other policies and regulations adopted by hospitals, clinics, and other healthcare organizations.
2. *Architectures for trusted HIS:* Development of formal, model-based design technology that integrates architecture modeling with privacy and security modeling for generating trusted HIS.
3. *Development of secure home-based health monitoring systems, integrated in the HIS of healthcare providers:* Consumer privacy protection recommendations for consumer adoption.

1A.4.1 Privacy Modeling and Analysis

We investigate two fundamental questions.

1. How does one describe formally and precisely privacy and security policies such that they faithfully capture the meaning of laws and regulations, clinical practices and provide foundation for formal reasoning and enforcement?

2. How does one extract formal operational rules (interactions, information flows, and restrictions) from audit logs of data access in a complex, dynamic clinical environment?

Privacy is an increasingly important business concern in health care, financial services, and other organizations. In this project, we build on our previous work to develop approaches for *modeling systems that handle sensitive information, languages for specifying privacy policies, and algorithms for their enforcement.* We also plan to develop some more extensive policy examples, such as a significant portion of HIPAA in a machine-processable format, and begin a new direction *incorporating risk management concepts into security analysis and decision making.* However, a significant amount of fundamental science remains to be done in each of these directions. Specifically, we plan to extend our privacy policy language, the Logic of Privacy and Utility (LPU), with internal operations (e.g., Alice uses Charlie's data to do X) so that we can look at polices for internal use within an organization. We also plan to extend LPU with data provenance concepts. Data provenance is relevant since regulations may not apply when data is publicly available. In addition to information about individuals, privacy policies often contain clauses that refer to aggregate information about a group of individuals. For example, the HIPAA Privacy Rule has a clause allowing *anonymized* protected health information to be released, where anonymized is defined to be as *certified by a qualified statistician.* In order to develop computational methods for specifying and enforcing such privacy properties, it is essential to make precise the notion of *data anonymity* or *data privacy* as used in such contexts. We propose to develop theoretically well-founded definitions for data privacy concepts, and integrate these concepts into our privacy policy language and enforcement methods.

Regarding the second question, our proposed research is connected to experimental efforts and rich data sets accessible at the Vanderbilt Medical Center. Our previous research has shown that the access logs of electronic medical records are essential part of policy enforcement. However, the study of the information within the access logs for health information privacy and security has been neglected. The size and detail, diversity, and complexity of work flows of electronic medical record systems access logs provides a prime opportunity to leverage automated methods to learn from, as well as monitor access logs. As opposed to having to model every workflowin the system, our automatic extraction techniques are going to speed up the process and provide models that are potentially much closer to reality.

1A.4.2 Architectures for Trusted HIS

The goal here is to develop a formalized, model-based design approach to trusted HIS. The proposed approach is based on the theoretical foundations and tools for model integrated computing (MIC) developed at Vanderbilt [36–39], platform-based design (PBD) advocated at UC Berkeley [40], and the results of the foundational work on privacy modeling and analysis (described above) at Stanford, CMU, and Vanderbilt. MIC is predicated on the notion that models play an essential role in every part of a system's life cycle, which commences with specification, progresses through design, development, verification, integration, and concludes with upgrade and maintenance.

PBD focuses on the creation of abstraction layers in the design flow, and investigates the semantic properties of mapping across these layers. To facilitate experimental work in a real-life, clinical setting and ease transitioning, we choose standard service-oriented architecture (SOA) [41] implementation as composition platforms to assemble loosely coupled software services characteristics for medical informatics applications. The combination of MIC, PBD, and SOA techniques can enable the design of complex HIS to ensure reliability, performance, privacy, and security beyond what can be achieved by current ad hoc practices. We answer the following questions:

1. How can we provide deep integration for models used in functional design and for privacy and security policies?
2. How can we create a run-time architecture that can be model driven and include components for enforcing dynamic security and privacy models?
3. How can we validate the proposed approach and show its effectiveness using relevant metrics?

The primary challenge in answering the first question stems from the fundamental difference between Domain Specific Modeling Languages (DSMLs) used in model-based design, and formal policy languages developed for modeling policies. DSMLs are usually defined using meta-models expressed in metamodeling languages [42–44]. Their semantics is frequently informal, or hidden in tools that check well-formedness of models. Policy modeling languages are XML-based (such as XACML and their semantics are not defined formally. The lack of precise semantics enables only shallow composition of the HIS models and security, privacy policies. Shallow composition means that functional design models (workflows, organization models, data models, deployment models, and so forth) are considered independent from policy models and influence system behavior via independent mechanisms (e.g., workflow engine for functional models and policy decision point, and policy enforcement points for policies). Furthermore, in the frame of dynamic systems where policies and functional models change independently and frequently, shallow composition is the source of errors and violations. Deep composition of functional modeling and policy modeling requires the development of common semantics foundation for functional and policy modeling languages. We have achieved significant progress in both areas. In model-based design, the typical solution used by meta-modeling tools is to annotate meta-models with a constraint language like OCL (Object Constraint Language). The Vanderbilt team has developed a theory for the structural semantics of DSMLs [45, 46]. Using this theory, meta-models are translated into constraints, so additional constraints can be injected directly into the resulting interpretation of the meta-model. We are extending this work in new directions on the separation of static and dynamic structural and policy constraints, where static constraints can be checked at design time, while dynamic constraints need to be checked at run time. We plan to develop consistency checking tools for resolving conflicts between functional models and policies, and work on the introduction of constructive modeling in the design flow [45] for repairing functional models that contradict policy requirements.

Answering the second question requires progress in extending standard SOA execution platforms (such as Apache ODE) with policy evaluation point that are Prolog based. We have already shown the feasibility of this approach [7–12]. However, significant further effort is required for the development of a suite of model transformation tools that can translate dynamic structural and policy constraints into Horn logic. We will build this effort on the MIC model transformation tool, GReAT [47] and FORMULA [48] that translates meta-models and models into Horn logic.

To answer the third question, we have designed and implemented the Model Integrated Clinical Information System (MICIS) platform [10–12] that serves as an evolving collaboration platform between TRUST researchers and the Vanderbilt Medical Center. The first real-life experiment using MICIS has been the sepsis treatment enhanced through electronic protocolization (STEEP). MICIS and its first major application STEEP is currently being tested and validated at the Vanderbilt Medical School Simulation Center and scheduled for clinical trials next year.

1A.4.3 Home-Based Health Monitoring

Recent advances in wireless sensor networks have made it possible to deploy wearable sensors on the bodies of patients in a residential setting, allowing continuous monitoring of physiological signals (such as ECG, blood oxygen levels), and other health-related information (such as physical activity levels) [49–66]. This provides a new technology option for reducing healthcare expenses through earlier detection of medical conditions. We address fundamental questions of scalability and extensibility; robust, reliable, and QoS-aware service delivery in a dynamic wireless environment; and privacy and security requirements for a medical practice. Our preliminary work has designed and built *CareNet*—a small-scale wireless sensor networking environment for remote sensing [41, 67, 68]. A significant amount of fundamental science remains to be done in each of the above directions. The medical sensing system that is built in the industrial, scientific and medical band suffers uncertainty in channel quality [69], user location [4], as well as coexisting devices and networks [34, 63]. Our previous work has developed an uncertainty-aware service admission control and robust routing algorithm based on the uncertainty-aware optimization framework [63, 64]. Privacy and security are primary considerations in our system design [70–72]. Our current system design features built-in secure communication components, which are adaptively implemented for different networking environments to protect the physiological information and video streams that carry the footage of the human movement [67]. As a next step, we will investigate detection mechanisms to identify anomaly situations caused by malicious attacks or device malfunctions. See [67] for some new directions.

1A.5 TRUST Research in Secure Physical Infrastructures

The third grand challenge thrust focuses on the development of science and technology for the security and improved functionality of the nation's physical infrastructure, including the power grid, telecom infrastructure, water transport system, and the interstate highway system. Since its inception, TRUST has developed wireless sensor networking technologies as a means for developing novel, powerful solutions for control, security, and maintenance [4, 67, 78]. We are particularly interested in the problems of connectivity/data delivery, security, robustness, and privacy in the adoption of wireless sensor networks. Over the next 5 years TRUST will emphasize research products for the power grid. Our general emphasis is on digital control systems (DCS) for SCADA systems—the overall collection of control systems that measure and change the infrastructure state to facilitate delivery of the commodity (electricity, water, gas, and oil). Wireless Sensor Network technology is one of the key areas of expertise of the TRUST faculty, with specific emphasis on closing the loop around these network-embedded systems. In contrast to a number of monitoring applications, this forces the development of hard real-time robust algorithms for control. Through partnerships with a number of industry leaders, we have established the viability of a large and vibrant industry sector in wireless sensor networks. Key to the development of this emerging industry sector is security and privacy solutions for DCS. Security is especially critical, given the possibility of some sensor nodes being appropriated by attackers. The development of the next generation of SCADA systems creates several network research challenges [85].

While supporting traditional requirements (such as scalability and efficiency) our research especially concentrates on providing reliable and secure operation, even if a subset of the network is controlled by an adversary or a malicious insider [86]. Given the distributed and unattended nature of SCADA devices, the possibility of compromised devices is a realistic threat.

1A.5.1 Scalable and Secure Routing

We are collaborating on the development of provably robust networking technologies for large sets of randomly distributed (and possibly malicious) sensors, building on and extending previous research by TRUST itself. We develop implementations of practical algorithms, which include means for identifying malicious nodes, using the past behavior of routers and other communication elements as the basis for establishing levels of trust. The key new science that is developing here relates to advances in multiplayer nonzero games with incomplete information. The games are between the attackers and the SCADA control schemes. Even the existence of solutions to multiplayer games with full information is nontrivial to establish, except in the case of Stackelberg or Nash strategies (see, for example, [87]). When there is partial collaboration among the players and incomplete information, the state of the art is rudimentary (see, for example, [88–90]). Furthermore, in multiple plays of the game, there is an opportunity for learning the utility functions of the players.

1A.5.2 Extending the Threat Models for Next Generation SCADA

Turning to the security issues, our first step in securing NG-SCADA has been the development of a security threat model [91]. The TRUST model has been extremely successful as the common basis for multi-institutional research. The model defines and explores the potential for mote-class attackers, laptop-class attackers, outsider attacks, and insider attacks. We develop this model to effectively categorize cyber attacks into taxonomies that illuminate the problem space, considering three distinct dimensions: how attacks manifest in appearance as seen in network traffic (defense perspective); how attacks are constructed, and the accompanying resources required to realize them (attacker and prevention perspective); and the damage implications of different types of attacks (adopting the victim's perspective). We also consider the problem of dealing with specific attacks on a SCADA system. Insider attacks and compromised nodes represent fundamental security risk in SCADA [92]. An insider can commit actions that lead to the physical damage of equipment; an attacker can remotely break into a SCADA master control station and cause catastrophic damage; a local attacker can compromise individual SCADA sensor node, inject malicious program code, tamper with data inputs, and so forth. The foundational new science to be developed here is an extension of the taxonomy of attacks and defenses from nontime-critical networks to hard real-time SCA-

DA networks, which are embedded in the physical infrastructure. Best-effort guarantees are inadequate for real-time SCADA networks, and consequently there are new classes of attacks, which rely on the time criticality of the data.

1A.5.3 Secure Control Systems

We propose reactive methods and algorithms that ensure secure control of SCADA network under previously considered attacks. We design control-theoretic algorithms that are resilient to deception and denial-of-service attacks [93–95]. While control theory has studied fault tolerance and robust control algorithms, there is no theory for the analysis and design of control algorithms for security. We also develop security architectures for control systems [82–84]. While fault-tolerant control architectures have previously incorporated redundancy and diversity; secure architectures need a new approach, where the interplay between redundancy, diversity, the principle of least privilege, and the principle of separation of duty are analyzed. We also propose new cryptographic protocols for the communications among entities, to prevent a single point of failure. Once again here the hard real-time issues combined with the need to close the loop (make decisions) generate a new class of security problems.

1A.5.4 Secure and Privacy Aware Demand Response Technology

Demand response systems use power consumption monitoring at the residential/small business level, coupled with short-term pricing information [95]. They are the key technology behind the development of smart grids, which will allow for decentralized generation sources (such as wind, solar, and other renewables). Another goal is the modification of customer electricity consumption during times of peak usage, in order to help address system reliability, reflect market conditions and pricing, and support infrastructure optimization or deferral. Yet another goal is the facilitation of demand shedding, the dynamic temporary reduction or curtailment of peak load when dispatched, and refers to strategies that can be possibly implemented within a shorter period of response time [96]. We demonstrate techniques for allowing customers to provide appliance usage information to an independent system operator's Web site for pricing information.

1A.5.5 Privacy and Monitoring Policy

Many of the above projects invoke specific privacy research problems. We are developing meaningful definitions/models for privacy in the context of shriveled infrastructure [97, 98]. We link sensor acuity and data access to mission, eliminating the overcollection of personal data. As part of this effort, we will develop privacy metrics for gauging the extent of personally identifiable information in collected data. We also develop data storage and query languages that protect personally identifiable information from third-party access, as well as from those that collect the information. We further propose to develop technologies that prevent data reuse and aggregation outside of mission requirements. One key to this effort will be the development of privacy-preserving, query-answering systems that answer queries while provably guaranteeing that sensitive information is kept secret. We have already derived the first provable privacy guarantee for the method of synthetic data generation, which is currently the state of the art in practical data privacy. We now propose adapting Digital Rights Management (DRM) technology to allow residents to maintain some level of control over their power consumption information.

1A.6 Shared Experimental Facilities

The hallmark of TRUST has been to establish a number of test beds, for integrating the research of the center and presenting it to stake holders for their evaluation. These test beds are described hereafter.

1A.6.1 The Cyber-Defense Technology Experimental Research Test Bed (DETER)

DETER was developed to provide a robust, large-scale test bed experimental infrastructure to support the development and demonstration of next-generation information security technologies. DETER provides a platform for secure, extensible, repeatable network and cyber security–related experimentation, and a means to validate theory and simulation. DETER provides an environment for testing hardware, software, tools, and techniques. Unlike simpler test beds constructed by manually wiring together and configuring a dedicated set of machines, DETER provides a general-purpose, shared, and remotely accessible facility for experimentation and testing. The test bed hardware can be partitioned into independent and isolated experimental workspaces, which can be used simultaneously and support multiple simultaneous experiments. DETER has been used by DHS to conduct such elaborate attack and defense scenarios, such as Cyberstorm in February 2007. Specific capabilities of DETER are listed hereafter.

- *Security:* DETER provides rigorous security mechanisms to protect not only the test bed and user applications, but also the public Internet. Security mechanisms address both infiltration and exfiltration, the protection of hardware and software, the isolation of experiments, and confidentiality and integrity of experiment data, source code, and results.
- *Containment:* DETER provides containment to support safe experiments that present a wide range of threat levels—from benign applications executing on a small number of nodes to *live* testing of a contagious attack program, whose attributes are completely unknown (e.g.,

an actual worm or virus). DETER also provides varying degrees of isolation and segregation of facilities, based on threat levels.

- *Fidelity:* DETER stresses the importance of fidelity to *real* networks, in particular the Internet, and achieves this with a large enough number of nodes, realistic router and end-system behavior, realistic heterogeneity of hardware and software, and a realistic mix of link bandwidth and delay.
- *Repeatability:* DETER provides a means of conducting repeatable experiments via test bed control software that collects and stores complete disk images, captures the operating system configuration, records the input/output state of an experiment, and logs relevant data.
- *Programmability:* DETER nodes are programmable, to allow users to conduct a variety of experiments without the need for vendors to open their platforms to experimental modifications.
- *Research functionality:* DETER was built to support research in security and, as such, the DETER team provides a technical and social environment for security researchers to share the results of their work, and expand their opportunities to collaborate.

The DETER system is somewhat detailed hereafter.

- *Equipment/space:* The DETER test bed consists of two clusters of experimental PC nodes, one at USC-ISI and another at UC Berkeley. Currently, there are nearly 500 nodes consisting of machines from Dell (860, 1850), Sun (V60x, V65x, X2100), and IBM (Netfinity, x330). These machines provide over 2,700 individual gigabit Ethernet ports. Both sites provide users with access to specialized hardware resources, such as some intrusion detection systems (IDSs), hardware routers, and field-programmable gate array (FPGA) boards.
- *Hybrid topologies:* DETER can dynamically create a number of topologies, with individual test bed nodes representing wired or wireless devices. Arbitrary topologies can be dynamically created and disassembled by SNMP agents via VLANs on the switches. Currently, DETER supports topologies consisting of wired network devices, as well as simulations of wireless network devices based on 802.11 (a/b/g), WiMAX, 3G, and wireless sensor network protocols.
- *Special hardware/software:* In addition to the core system nodes, the DETER test bed includes a number of specialized hardware and software components that allow users to test and evaluate their code with a number of commercial hardware devices, and with a number of commer-

cial software applications. Currently, the following are available: Cloud Shield 2200, HP DL360G2 Minibed, HP Procurve router, Juniper M7i routers, Juniper IDP-200 IDS, McAfee Intrushield 2600, and NetFPGA2 boards. Additionally, DETER nodes provide for experimentation in the following operating environments: Windows XP/Server 2003, RedHat Linux 9/Fedora Core4, and FreeBSD 4.10/5.4/6.1.

- *Scale:* The DETER test bed supports experimentation with both physical hardware devices and emulated devices. It also supports emulation to allow the execution of large experiments, especially those requiring many more nodes than can be provided physically. Emulation has enabled researchers to conduct experiments on code, such as the Slammer worm in an environment representing a 1/64th scale version of the Internet.
- *Automation:* The DETER test bed provides a set of common software tools and data that support security experiments. These include building blocks, data sets, a repository of complete experiments, and an integrated experimental environment. The tools support measurement, data analysis, and visualization, including generators for sample topologies, network configurations, and attack and background traffic. Many of these tools were initially developed by DETER experimenters, and the data sets contain standard topologies, and other static information that facilitate comparative experiments using common network conditions, while the repository of complete experiments contains samples ranging from very simple to complex.

1A.6.2 Command and Control (C2) Wind Tunnel (C2W)

C2W is high-fidelity, multimodel simulation environment for integrating, testing, and validating large-scale, software intensive, networked mission systems. The mission systems include C2 organization models, C2 technical architecture models, network models, platform models, senor models, physical environment models, and experiment models. C2W supports computational experimentation using virtual component, complemented with component-in-the-loop simulations where real components are integrated with simulations. The architecture enables the integration with components deployed on emulated platforms, such as EMULAB. With increased attention focused on cyber warfare, especially highlighted by the cyber attacks in Estonia and recently in Georgia, it has proved to be important to be able to go beyond DETER to be able to simulate command and control of cyber and physical infrastructures. C2W represents such an effort and is strongly used by the Air Force for modeling cyber operations jointly with DHS and the newly created Joint Forces Cyber Command. C2W includes a simulation component integration framework that is based on the high level architecture (HLA), and a model integration framework that is based on the model integrated computing infrastructure at Vanderbilt. Specification of computational experiments is fully model based; the models of the mission system, mission scenario and experiments are defined and the system is *autointegrated from the models.*

The currently available simulation components include: Colored Petri Net (CPN) Tools (to model and simulate organizations and decision making processes); Simulink/Stateflow (to model and simulate embedded controllers, vehicle dynamics); OmNet++ (to model and simulate local area, wide area, and wireless networks); Delta-3D (modeling environment and physics); DeVS (modeling software components) and emerging model libraries for simulating network attacks. The infrastructure is built using standard interfaces, and can include open source and proprietary implementations. C2W provides the software infrastructure for modeling and generating large scale experiments. Integrating the component integration platform with DETER provides the opportunity for generating precise simulations for complex, distributed missions, and for investigating the impact of cyber attacks on mission performance. The full tool suite provides a test range for technologies developed for increasing systems resilience against a range of network and other cyber attacks. The test bed is widely used by industry and government partners.

1A.6.3 Experimental Social Science Laboratory (Xlab)

Xlab is a Berkeley laboratory for conducting experiment-based investigations of issues of interest to social scientists. The Xlab was established in 2004 with a grant from the National Science Foundation for research in the emerging field of experimental economics, pioneered by Nobel Laureate Ackerlof. It is currently part of the Institute of Business and Economic Research, and is supported by the Haas School of Business at Berkeley, the California Management Review, and Hewlett Packard Labs (see http://xlab.berleley.edu). Xlab provides an environment in which controlled laboratory experiments can be performed in areas such as experimental economics, usability, and other social science fields that, until now, have not made much use of experimentation. On TRUST the major use of Xlab has been to do privacy studies in a controlled environment, and to perform experiments on the economic incentives for use of new password protection schemes and other encryption schemes. It is a key component of the privacy and usability studies for trustworthy systems on TRUST.

1A.6.4 SCADA Test Beds

Supervisory Control and Data Acquisition (SCADA) refers to a large-scale, distributed measurement (and control) system. SCADA systems are used to monitor, or to control chemical or transport processes, in municipal water supply systems, to control electric power generation, transmission and distribution, gas and oil pipelines, and other distributed processes. Given the critical nature of the SCADA systems, ensuring their security is of great importance. We have developed a SCADA reference architecture based on a layered model, ranging from the low-level sensors and actuators, through the remote terminal units (RTUs), to the SCADA hosts, and connected up to the corporate network. The corporate network represents the business end of a utility. This network is typical of an enterprise with a LAN/WAN connected to the Internet. However, in the case of utilities and industrial plants, the corporate network is often connected to the SCADA network, in order to simplify business processes by allowing network access to critical data on SCADA servers. We have thus far prototyped and instantiated the above architecture, within an experimental setup that serves as a shared experimental platform for researchers inside and outside of TRUST. Some further refinement and developments will be made in the following years of the project. The proposed SCADA framework integrates prototype SCADA computing nodes, a real-time plant simulator, and an emulated network infrastructure where security attacks could be studied. The SCADA computing nodes implement low-level (process) regulators, data acquisition functions, as well as higher-level supervisory control functions. These nodes are connected to the real-time plant simulator via the same real-time interfaces as used in physical plants, and they are also networked via a network emulation system that allows the tightly controlled and carefully monitored execution of cyber attacks. For the latter we propose to use the DETER (or a DETER-like) test bed that runs these attacks on an isolated, yet realistic network. We use a model-based integration and design environment for the design, execution, and analysis of experiments on the test bed.

Demand Response SCADA Test Bed

The goal of this test bed effort is for testing advanced demand response technologies, which are extremely important for emerging smart grid applications, especially those involving distributed supply and demand. The emphasis on secure SCADA for these applications rely on the development of Intrusion Detection System (IDS) technology that can: (1) efficiently detect and block cyber intrusions into SCADA systems in real operational environments, and in real time; (2) without interrupting the control performance of the protected system; (3) without creating extra operational burden or operational reservations due to false alarms; and (4) in the presence of both malicious and messily benign network traffic. The system must operate in a real-time, robust fashion, with performance adequate to meet the demands of the dynamic cyber-physical interactions inherent to SCADA systems. We are constructing a test environment, consisting of physical programmable logic controllers (PLCs) and intelligent electronic devices (IEDs), to emulate the SCADA system under study. We plan to inject designed attack traffic, along with traffic synthesized from traces separate from those used in developing and tuning the system in order to assess false positive and false negative rates. We also leverage malware defense mechanisms proposed by industry (such as TCG's TPM chips or ARM's TrustZone), as well as software-based attestation technology, recently developed by TRUST itself [99]. These technologies enable a remote verifier to check the correctness of software executing on SCADA devices. The demand response test bed emulate online, reporting of power generation and consumption information, to support the ongoing focus on the privacy risks arising from the collection of power consumption data in current and future demand-response systems [100, 101].

1A.6.5 Home Monitoring Test Bed

As stated in Section 1A.4.3, recent advances in wireless sensor networks have made it possible to deploy wearable sensors on the bodies of patients in their residential setting. In order to explore this unique opportunity, TRUST sponsored research in the past 5 years, devoted to design, development, and test of a prototype system CareNet—an integrated wireless sensor networking environment for remote healthcare—through a collaborative effort between Cornell, Berkeley, and Vanderbilt. The goal of this remote healthcare system is to provide comprehensive patient monitoring services with security, privacy, performance and reliability assurance, and real-time medical data access via an interface with the electronic medical record system.

As shown in Figure 1A.1, the CareNet system has three major components: (1) medical sensing subsystem, which provides reliable and QoS-aware networking of heterogeneous sensors, and privacy-preserving medical data collection; (2) data analysis and transport, which analyzes the data for timely

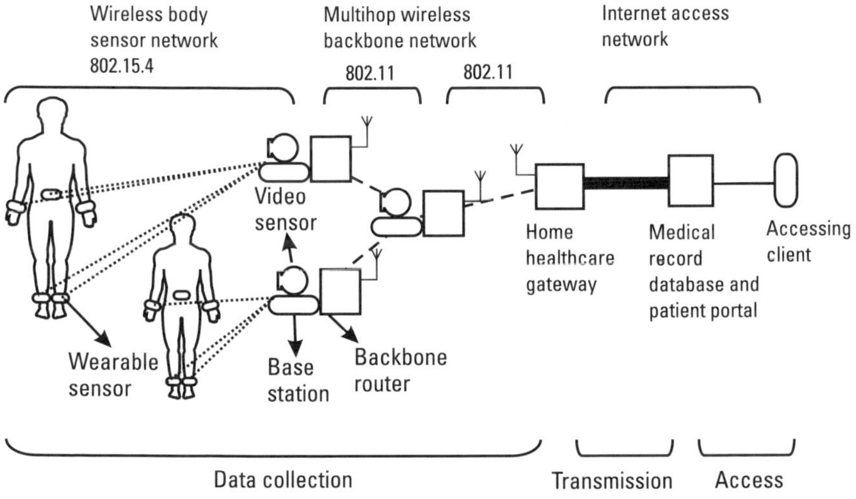

Figure 1A.1 Graphical representation of the CareNet system.

detection of critical medical conditions, and provides secure remote communication between the patient home and the care provider; and (3) EMR/patient portal interface, which provides medical data archive and interface for accessing the monitoring results. Our research and development effort on CareNet is focused on the points detailed hereafter:

- Privacy and security;
- Reliability and QoS awareness;
- Integration with the Web-based patient portal.

1A.7 Concluding Remarks

TRUST has set ambitious goals research, education, and knowledge transfer, to position itself as a leading authority and developer of a science base for cyber security. At the same time, TRUST will continue to educate the next generation of researchers and experts, so that they are well armed to address the technical, policy, and economic issues, through increased capacity for interdisciplinary collaborations, essential to improving cyber security and building trustworthy systems, as well as expanding partnerships, collaboration with industry, and international affiliations to be in the center of the action. To that end, the TRUST research agenda provides students and faculty the opportunity to engage in crossinstitution and interdisciplinary activities that are only possible with the intellectual and financial support of the Center. These research activities each include education components focused on producing leading cyber security teaching material for the classroom, professional training, or other learning opportunities, which are also be disseminated widely beyond TRUST. Outreach and diversity activities by TRUST produce a more diverse science and engineering workforce, to meet the future demands of our technical profession. TRUST infuses the engineering pipeline with new, diverse, and talented individuals, while preparing the next generation of cyber security professionals for successful careers. TRUST is also expanding the relatively small cadre of legal and economic researchers engaged in interdisciplinary cyber security research by including postdocs and Ph.D.s in noncomputer science and engineering disciplines. The legacy of TRUST is the creation of a science base for security, an education base, and a technology and policy transfer methodology for cyber security. The long-term research agenda of TRUST will not only advance the frontiers of knowledge in trustworthy computing, but will also influence, on a national and also international scale, future academic research, industrial research and development, policy advocacy, and education. This progress will occur as TRUST researchers identify new directions of inquiry, disseminate their findings to the broader community, and produce a culturally diverse group of professionals, steeped in interdisciplinary approaches to research, with the result of further advancement of the field of trustworthy systems.

Acknowledgments

This work was supported in part by TRUST (Team for Research in Ubiquitous Secure Technology), which receives support from the National Science Foundation (NSF award number CCF-0424422) and the following organizations: British Telecom, Cisco, DoCoMo-USA Labs, EADS, ESCHER, HP,IBM, iCAST, Intel, Microsoft, ORNL, Pirelli, Qualcomm, Sun, Symantec, Tata Consultancy Services, Telecom Italia, and United Technologies.

References

[1] Dhamija, R., and J. D. Tygar, "Dynamic Security Skins," in *Phishing and Countermeasures: Understanding the Increasing Problem of Electronic Identity Theft*, S. Myers and M. Jakobsson, (eds.), New York: Wiley-Interscience, 2007, pp. 339–351.

[2] Dhamija, R., and J. D. Tygar, "The Battle Against Phishing: Dynamic Security Skins," *SOUPS 2005: Proceedings of the 2005 ACM Symposium on Usable Security and Privacy*, July 2005, pp. 77–88.

[3] "For the Record: 'Protecting Electronic Health Information,'" Committee on Maintaining Privacy and Security in Health Care Applications of the National Information Infrastructure, NRC-CSTB, National Academy Press, Washington, D.C., 1997.

[4] Picker Institute, "Eye on Patients, American Hospital Association and the Picker Institute," 2000.

[5] Masys, D., "The Informatics of 21st Century Healthcare," *Workshop on Model-Based Design of Trustworthy Health Information Systems*, Nashville, TN, September 30, 2007.

[6] Grossman, J. H., "Disruptive Innovation in Health Care: Challenges for Engineering," *The Bridge—Publication of National Academy of Engineering*, Vol. 38, No. 1, 2008.

[7] Mathe, J. L., et al., "Towards the Security and Privacy Analysis of Patient Portals," *ACM SIGBED Review*, Vol. 4, No. 2, April 2007.

[8] Werner, J., et al., "Platform-Based Design for Clinical Information Systems," *Proc. of the 5th IEEE International Conference on Industrial Informatics (INDIN 2007)*, Vol. 2, Vienna, Austria, June 2007, pp. 749–754.

[9] Mathe, J., et al., "Implementing a Model-Based Design Environment for Clinical Information Systems," *Workshop on Model-Based Trustworthy Health Information Systems (MOTHIS)*, Nashville, TN, September 2007.

[10] Duncavage, S., et al., "A Modeling Environment for Patient Portals," *Proc. Symposium of American Medical Informatics Association (AMIA)*, November 2007.

[11] Brantley, J. M., et al., "A Model-Integrated Approach to Implementing Individualized Patient Care Plans Based on Guideline-Driven Clinical Decision Support and Process

Management," *2nd International Workshop on Model-Based Design of Trustworthy Health Information Systems (MOTHIS 2008)*, Toulouse, France, October 1, 2008.

[12] Mathe, J., et al., "Model-Based Design of Clinical Information Systems," *Methods of Information in Medicine*, Vol. 47, No. 5, 2008, pp. 399–408.

[13] Brantley, J. M., et al., "A Model-Integrated Approach to Implementing Individualized Patient Care Plans Based on Guideline-Driven Clinical Decision Support and Process Management," *2nd International Workshop on Model-Based Design of Trustworthy Health Information Systems (MOTHIS 2008)*, Toulouse, France, October 1, 2008.

[14] Clay, W., "Botnets, Cybercrime and Terrorism: Vulnerabilities and Policy Issues for Congress," *Congressional Research Service*, updated January 29, 2008.

[15] Zhuang, L., et al., "Characterizing Botnets from Email Spam Records," *Proceedings of First USENIX Workshop on Large Scale Exploits and Emergent Threats (LEET)*, 2008.

[16] Karlof, C., J. D. Tygar, and D. Wagner, "A User Study Design for Comparing the Security of Registration Protocols," *Proceedings 1st USENIX Workshop on Usability, Psychology, and Security (UPSEC)*, April 2008.

[17] Karlof, C., et al., "Dynamic Pharming Attacks and Locked Same-Origin Policies for Web Browsers," *Proceedings of the Fourteenth ACM Conference on Computer and Communications Security (CCS 2007)*, October 2007, pp. 58–71.

[18] Karlof, C., and U. Shankar, A *Usability Study of Doppelganger, a Tool for Better Browser Privacy*, Technical Report UCB/EECS-2007-116, University of California at Berkeley, September 2007.

[19] Shankar, U., and C. Karlof, "Doppelganger: Better Browser Privacy Without the Bother," *Proceedings of the Thirteenth ACM Conference on Computer and Communications Security*, November 2006, pp. 154–167.

[20] Dhamija, R., J. D. Tygar, and M. Hearst, "Why Phishing Works," *Proceedings of CHI-2006: Conference on Human Factors in Computing Systems,* April 2006.

[21] Barreno, M., et al., "Open Problems in the Security of Learning," *Proceedings of the First ACM Workshop on AISec.*, 2008.

[22] Benjamin, I. P., et al., "Evading Anomaly Detection Through Variance Injection Attacks on PCA," *Proceedings of the 11th International Symposium on Recent Advances in Intrusion Detection (RAID 2008)*, 2008.

[23] Benjamin, I. P., et al., *Compromising PCA-Based Anomaly Detectors for Network-Wide Traffic*, UC Berkeley, Department of EECS Technical Report UCB/EECS-2008-73, May 29, 2008.

[24] Barreno, M., et al., *The Security of Machine Learning*, UC Berkeley, Department of EECS Technical Report UCB/EECS-2008-43, April 24, 2008.

[25] Nelson, B., et al., "Exploiting Machine Learning to Subvert Your Spam Filter," *Proceedings USENIX Workshop on Large-Scale Exploits and Emergent Threats (LEET)*, April 2008.

[26] Nelson, B., and A. D. Joseph, "Bounding an Attack's Complexity for a Simple Learning Model," *Proceedings of the First Workshop on Tackling Computer Systems Problems with Machine Learning Techniques (SysML)*, Saint-Malo, France, June 2006.

[27] Barreno, M., et al., "Can Machine Learning Be Secure?" *Proceedings of the ACM Symposium on Information, Computer and Communications Security (ASIACCS'06)*, Taipei, Taiwan, March 2006.

[28] Zhuang, L., F. Zhou, and J. D. Tygar, "Keyboard Acoustic Emanations Revisited," *Proceedings of the 12th ACM Conference on Computer and Communications Security*, November 2005.

[29] Zhuang, L., F. Zhou, and J. D. Tygar, "Case Study: Acoustic Keyboard Emanations," in *Phishing and Countermeasures: Understanding the Increasing Problem of Electronic Identity Theft*, M. Jakobsson and S. Myers, (eds.), New York: Wiley-Interscience, 2007, pp. 221–240.

[30] Lee, D., S. Shieh, and J. D. Tygar, *Computer Security in the 21st Century*, New York: Springer, 2005.

[31] Perrig, A., and J. D. Tygar, *Secure Broadcast Communication in Wired and Wireless Networks*, New York: Springer (Kluwer), 2003.

[32] Tygar, J. D., "Technological Dimensions of Privacy in Asia," *Asia-Pacific Review*, Vol. 10, No. 2, November 2003, pp. 120–145.

[33] Tygar, J. D., "Privacy in Sensor Webs and Distributed Information Systems," in *Software Security*, B. Pierce, et al., (eds.), New York: Springer, 2003, pp. 84–95.

[34] Hong, J., et al., "Privacy and Security in the Location-Enhanced World Wide Web," *Proceedings of the Workshop on Privacy at Ubicomp 2003*, October 2003.

[35] Tygar, J. D., "The Problem with Privacy," *Proceedings of the 2003 IEEE Workshop on Internet Applications*, June 2003, pp. 2–8.

[36] Sztipanovits, J., and G. Karsai, "Model-Integrated Computing," *IEEE Computer*, Vol. 30, April 1997, pp. 110–112.

[37] Karsai, G., et al., "Model-Integrated Development of Embedded Software," *Proceedings of the IEEE*, Vol. 91, No. 1, January 2003, pp. 145–164.

[38] Karsai, G., et al., "Towards Two-Level Formal Modeling of Computer-Based Systems," *Journal of Universal Computer Science*, Vol. 6, No. 11, November 2000, pp. 1131–1144.

[39] Ledeczi, A., et al., "Composing Domain-Specific Design Environments," *IEEE Computer*, November 2001, pp. 44–51.

[40] Sangiovanni-Vincentelli, A., "Defining Platform-Based Design," *EEDesign Exclusive Feature*, March 16, 2002.

[41] Breu, R., J. Sztipanovits, and E. Ammenwerth, "Model-Based Design of Trustworthy Health Information Systems," *Methods of Information in Medicine*, Vol. 47, No. 5, 2008, pp. 389–391.

[42] Karsai, G., et al., "Model-Integrated Development of Embedded Software," *Proceedings of the IEEE*, Vol. 91, No. 1, January 2003, pp. 145–164.

[43] Karsai, G., et al., "Towards Two-Level Formal Modeling of Computer-Based Systems," *Journal of Universal Computer Science*, Vol. 6, No. 11, November 2000, pp. 1131–1144.

[44] Ledeczi, A., et al., "Composing Domain-Specific Design Environments," *IEEE Computer*, November 2001, pp. 44–51.

[45] Jackson, E., and J. Sztipanovits, "Constructive Techniques for Meta- and Model-Level Reasoning," *Model Driven Engineering Languages and Systems (MODELS)*, October 2007, pp. 405–419.

[46] Jackson, E. K., and J. Sztipanovits, "Towards a Formal Foundation for Domain Specific Modeling Languages," *Proc. 6th ACM International Conference on Embedded Software (EMSOFT'06)*, Seoul, South Korea, October 2006.

[47] Karsai, G., et al., "On the Use of Graph Transformation in the Formal Specification of Model Interpreters," *Journal of Universal Computer Science*, Vol. 9, No. 11, November 2003, pp. 1296–1321.

[48] Jackson, E., W. Schulte, and J. Sztipanovits, *The Power of Rich Syntax for Model-Based Development*, Technical Report MSR-TR-2008-86, Microsoft Research, Redmond, WA, June 2008.

[49] Giorgetti, G., et al., "The Personal Sensor Network: A User-Centric Monitoring Solution," *Proc. of BodyNets*, 2007.

[50] Dabiri, F., H. Noshadi, and H. Hagopian, "Lightweight Medical BodyNets," *Proc. of BodyNets*, 2007.

[51] Al-Omari, S., and W. Shi, "Toward Highly-Available WSNs for Assisted Living," *Proc. of HealthNet*, 2007.

[52] Natarajan, A., et al., "Investigating Network Architectures for Body Sensor Networks," *Proc. of HealthNet*, 2007

[53] Ruzzelli, A. G., et al., "Energy-Efficient Multi-Hop Medical Sensor Networking," *Proc. of HealthNet*, 2007.

[54] Sammon, M. J., et al., "MACCS: An Industrial Study of Hands-Free Wireless Communications for Mobile Healthcare Workers," *Proc. of HealthNet*, 2007.

[55] Gao, T., et al., "Participatory User Centered Design Techniques for a Large Scale Ad-Hoc Health Information System," *Proc. of HealthNet*, 2007.

[56] Lorincz, K., et al., "Sensor Networks for Emergency Response: Challenges and Opportunities," *IEEE Pervasive Computing, Special Issue on Pervasive Computing for First Response*, 2004.

[57] Shnayder, V., et al., *Sensor Networks for Medical Care*, Harvard University Technical Report TR-08-05, 2005.

[58] Gao, T., et al., "Vital Signs Monitoring and Patient Tracking over a Wireless Network," *Proc. of the 27th IEEE EMBS Annual International Conference*, 2005.

[59] Gao, T., D. Greenspan, and M. Welsh, "Improving Patient Monitoring and Tracking in Emergency Response," *Proc. of the Intl. Conference on Information Communication Technologies in Health*, 2005.

[60] Wood, A., et al., *ALARM-NET: Wireless Sensor Networks for Assisted-Living and Residential Monitoring*, University of Virginia Technical Report CS-2006-11, 2006.

[61] Virone, G., et al., "An Assisted Living Oriented Information System Based on a Residential Wireless Sensor Network," *Proc. of Distributed Diagnosis and Home Healthcare (D2H2)*, 2006.

[62] Stankovic, J. A., et al., "Wireless Sensor Networks for In-Home Healthcare: Potential and Challenges," *Proc. of HCMDSS Workshop*, 2005.

[63] Hou, J., "PAS: A Wireless-Enabled, Sensor-Integrated Personal Assistance System for Independent and Assisted Living," *Proc. of Joint Workshop on HCMDSS and MDPnP Interoperability*, 2007.

[64] Wang, Q., et al., "I-Living: An Open System Architecture for Assisted Living," *Proc. IEEE SMC*, 2006.

[65] Otto, C., et al., "System Architecture of a Wireless Body Area Sensor Network for Ubiquitous Health Monitoring," *Journal of Mobile Multimedia*, Vol. 1, No. 4, 2006.

[66] Milenkovic, A., C. Otto, and E. Jovanov, "Wireless Sensor Networks for Personal Health Monitoring: Issues and an Implementation," *Journal of Computer Communications*, 2006.

[67] Jiang, S., et al., "CareNet: An Integrated Wireless Sensor Networking Environment for Remote Healthcare," *Proc. of Intl. Conf. on Body Area Networks (BodyNet)*, Tempe, AZ, March 2008.

[68] Kuryloski, P., et al., "MedSN System for In-Home Patient Monitoring: Architecture, Privacy and Security," *Proc. of the Joint Conference on High Confidence Medical Devices, Software, and Systems and Medical Device Plug-and-Play Interoperability*, June 2007.

[69] Cerpa, A., et al., "Temporal Properties of Low Power Wireless Links: Modeling and Implications on Multi-Hop Routing," *Proc of ACM MobiHoc*, 2005.

[70] Wu, T., Y. Xue, and Y. Cui, "Privacy Preservation in Wireless Mesh Network," *Security in Wireless Mesh Networks*, Boca Raton, FL: Auerbach Publication, 2008.

[71] Wu, T., et al., "A Fast and Efficient Source Authentication Solution for Broadcasting in Wireless Sensor Networks," *Proc. of New Technologies, Mobility and Security*, May 2007.

[72] Wu, T., Y. Cui, and Y. Xue, "Preserving Traffic Privacy in Wireless Mesh Networks," *Proc. of IEEE International Symposium on a World of Wireless, Mobile and Multimedia Networks (WoWMoM)*, Buffalo, NY, June 2006.

[73] Dai, L., et al., "Integrating Traffic Estimation and Routing Optimization for Multi-Radio Multi-Channel Wireless Mesh Networks," *Proc. of IEEE INFOCOM*, Phoenix, AZ, April 2008.

[74] Dai, L., et al., "Optimal Routing for Wireless Mesh Networks with Dynamic Traffic Demand," *Mobile Networks and Applications, Special Issue on Advances in Wireless Mesh Networks*, 2008.

[75] Wellons, J., et al., "Predictive or Oblivious: A Comparative Study of Routing Strategies for Wireless Mesh Networks Under Uncertain Demand," *Proc. of IEEE Communications Society Conference on Sensor, Mesh and Ad Hoc Communications and Networks (SECON)*, San Francisco, CA, June 2008.

[76] Wellons, J., and Y. Xue, "Oblivious Routing for Wireless Mesh Networks," *Proc. of IEEE International Conference on Communications (ICC)*, Beijing, China, May 2008.

[77] Dai, L., et al., "Throughput Optimization Routing Under Uncertain Demand for Wireless Mesh Networks," *Proc. of IEEE International Conference on Mobile Ad-Hoc and Sensor Systems (MASS)*, Pisa, Italy, October 2007, pp. 1–11.

[78] Masys, D., *Electronic Medical Records and Secure Patient Portals,* Technical report, Vanderbilt University Medical Center, October, 2005, White Paper from Design Workshop for an Integrative Project Related to Patient Portals, Vanderbilt University, Nashville, TN, December 16, 2005.

[79] Gonzalez, M. C., C. A. Hidalgo, and A. -L. Barabasi, "Understanding Individual Human Mobility Patterns," *Nature,* 2008.

[80] Chen, J., et al., "Wearable Sensors for Reliable Fall Detection," *Proc. of IEEE Engineering in Medicine and Biology Society,* 2005.

[81] Xue, Y., Y. Cui, and K. Nahrstedt, "Maximizing Lifetime for Data Aggregation in Wireless Sensor Networks," *Mobile Networks and Applications (MONET),* Special Issue on Energy Constraints and Lifetime Performance in Wireless Sensor Networks, Vol. 10, No. 6, December 2005, pp. 853–864.

[82] Jiang, S., Y. Xue, and D. Schmidt, "Minimum Disruption Service Composition and Recovery over Mobile Ad Hoc Networks," *Computer Networks,* Vol. 53, No. 10, July 14, 2009, pp. 1649–1665.

[83] Wu, T., et al., "Digital Rights Management for Video Sensor Network," *Proc. of IEEE International Symposium on Multimedia (ISM),* San Diego, CA, December 2006.

[84] Jiang, S., Y. Xue, and D. Schmidt, "Minimum Disruption Service Composition and Recovery over Mobile Ad Hoc Networks," *Proc. of the 4th Annual International Conference on Mobile and Ubiquitous Systems: Computing, Networking and Services (MOBIQUITOUS),* Philadelphia, PA, August 2007.

[85] Sastry, S., J. Sztipanovits, and H. Gill, "New Vistas in CIP Research and Development: Secure Network Embedded Systems," *Report of the NSF/OSTP Workshop on Innovative Information Technologies for Critical Infrastructure Protection,* September 19–20, 2002.

[86] Krogh, B., M. Ilic, and S. Sastry, *National Workshop on Beyond SCADA: Networked Embedded Control for Cyber-Physical Systems (NEC4CPS): Research Strategies and Roadmap,* Annual Report, 2007.

[87] Bohacek, S., et al., "Game Theoretic Stochastic Routing for Fault Tolerance on Computer Networks," *IEEE Transactions on Parallel and Distributed Systems,* Vol. 18, 2007, pp. 1227–1240.

[88] Cardenas, A. A., et al., "Cybersecurity: Basic Defenses and Attack Trends," *Homeland Security Technology Challenges: From Sensing and Encrypting to Mining and Modeling,* G. Franceschetti and M. Grossi, (eds.), Norwood, MA: Artech House, 2008.

[89] Cardenas, A. A., S. Amin, and S. Sastry, "Research Challenges for the Security of Control Systems," *3rd Usenix Workshop on Hot Topics in Security (HotSec 2008), associated with 17th Usenix Security Symposium,* San Jose, CA, July 2008.

[90] Barreno, M., A. Cardenas, and D. Tygar, "Optimal ROC Curves for a Combination of Classifiers," *Neural Information Processing Systems (NIPS) Proceedings,* Cambridge, MA: MIT Press, 2008, pp. 57–64.

[91] Roosta, T., et al., "The Inherent Security of Routing Protocols in Ad-Hoc and Sensor Networks," *Proceedings of the 50th Annual IEEE Global Communications Conference (IEEE GLOBECOM'07),* Washington, D.C., November 26–30, 2007.

[92] Roosta, T., S. Shieh, and S. Sastry, "Taxonomy of Security Attacks in Sensor Networks," *2006 American Control Conference*, Minneapolis, MN, 2006.

[93] Federal Technical Support Working Group (TSWG), "Sustainable Security for Infrastructure SCADA," http://www.tswg.gov/subgroups/ps/infrastructure-protection/documents/SustainableSecurity.pdf.

[94] Sinopoli, B., et al., "Distributed Control Applications Within Sensor Networks," *Proceedings of the IEEE, Special Issue on Sensor Networks and Applications*, August 2003.

[95] Sinopoli, B., et al., "Kalman Filtering with Intermittent Observations," *IEEE Transactions on Automatic Control, Special Issue on Networked Control Systems*, September 2004

[96] Schenato, L., et al., "Foundations of Control and Estimation over Lossy Networks," *Proceedings of the IEEE, Special Issue on Networked Control Systems*, January 2007.

[97] "Demand Response: Passive Proximity Electric Sensing," *Berkeley Sensor & Actuator Center (BSAC) Demand Response Enabling Technologies Development Workshop*, June 10, 2004.

[98] Bennett, C., S. B. Wicker, and J. Cardell, "Residential Demand Response Wireless Sensor Network," *Fourth Annual Carnegie Mellon Conference on the Electricity Industry—Future Energy Systems: Efficiency, Security, Control*, March 2008.

[99] Seshadri, A., et al., "SWATT: SoftWare-Based Attestation for Embedded Devices," *Proceedings of the IEEE Symposium on Security and Privacy*, Oakland, CA, May 2004.

[100] "Demand Response: Passive Proximity Electric Sensing," *EECS Department and the Berkeley Sensor & Actuator Center (BSAC) Demand Response Enabling Technologies Development Workshop*, June 10, 2004.

[101] Wagner, D., and D. Mulligan, "Network Security Architecture for Demand Response Sensor Networks," prepared for the California Energy Commission by CyberKnowledge and UC Berkeley, October 2005.

2

Homeland Security and National Defense in the Twenty-First Century

Pietro Finocchio and Pompeo Piedimonte

> There can be no freedom without security.
> —Wilhelm von Humboldt

> The dinosaurs disappeared because they relied on cuirasses rather than intelligence. They could not adapt to their changing environment. We shall disappear if we cannot adapt to an environment that now contains spaceships, computers and thermonuclear weapons.
> —Sir Arthur Clarke

2.1 The New World Scenario

The conclusion of the Cold War marked the end of a geopolitical scenario dominated by a bipolar balance of power in which only two superpowers, the United States and the Soviet Union, faced each other on a global scale due to their dominance in the politic, economic, industrial and technological fields. Therefore, the two countries were able to field a full range of military capabilities, to project and support them anywhere in the world, in order to exercise their economic and political influence over their satellite nations.

After the end of that era, paradoxically stable and universally known as the era of the iron curtain, which ended thanks to the fall of the Berlin Wall, the global political landscape has evolved towards a new scenario of regional balances of powers, which escalated into a series of local tensions and crises. Moreover, the attacks to the Twin Towers in New York, Atocha Station in Madrid, London's subway, and, more recently, the Taj Mahal Hotel in Mumbai show a qualmless threat without geographical limits. Consequently, there was and still is a high probability of asymmetric conflict and attacks, in and out homelands, which sometimes affect mainly defenseless civil populations, in which strategies, tactics and nonconventional means might be used by several entities, including nongovernmental groups, such as terrorists or extremists, that deliberately operate outside the rules of a morally acceptable behavior, as the standard uses of the western world are not necessarily shared, in order to exploit the effect of media and public sensitivity towards the loss of lives. The terrorists, with little or no respect for humanitarian law and common ethics, direct their attacks against both civilian and military targets; therefore this kind of threat requires addressing the issue of safety in a dual perspective (civil and military) both in homeland and in outside areas of operations as already mentioned before. In addition to this change, the growing dematerialization of the economy, the excess of production capacity and the saturation of the market consumer, the appearance of new superpowers are inducing economic, social, and political changes in more advanced countries of the so-called Western world.

In this new political landscape Italy, a long-time member of NATO, has a strategic geopolitical position, being located in the center of the Mediterranean, between Europe and Africa, the Middle East, and the Atlantic. Italy is, furthermore, a nation sensitive to religious issues having in Rome:

- The Vatican, which is the center of Catholicism;
- The oldest Jewish community outside Israel;
- The largest and most prestigious mosque in Europe.

In addition, Italy is subject to a heavy immigration of people who practice other religions. Over the millennia Italy had relationships with the other peoples of the enlarged Mediterranean basin and have been subject to foreign dominations, before unification; all this allowed the development of a multiethnic nature and culture (even if not consciously). Therefore, because of its geopolitical position and the other reasons already expressed, Italy can be considered archetypal for many other nations.

2.2 National Defense—Homeland Protection

In this new geopolitical landscape, armed forces have been compelled to assume a different role and the concept of national defense has evolved towards the one of homeland protection. Homeland protection implies the protection of real critical infrastructures (such as electricity, transportation, banking, and telecom) and virtual structures (such as cyber transactions) while preserving the privacy of the citizens, the national sovereignty, and, in general, national interest and welfare wherever they are threatened. Moreover, homeland security should include not only people and land of the nation, but also its culture, traditions, and ideal accomplishments that have flourished over the centuries.

Given this definition, homeland protection is in general conceived in most nations as composed by two branches: military defense (homeland defense) and civilian defense (homeland security). Both branches have specific characteristics and duties, but the cooperation and coordination between military defense and civilian defense is crucial in the case of large events inside the country (natural or anthropic catastrophic events, events that involve many authorities and people) or civil military cooperation (CIMIC) in military operations inside or outside the country. Following this direction, and compelled by national budgets restraints, in order to optimize the available resources, some nations (Italy has been precursor in this) are developing dual-use systems to be used for both military and civilian defense.

One can expect a near future when the majority of nations will adopt this approach. More specifically, the armed forces in general can contribute to the homeland security in the following areas:

- Intelligence capabilities;
- Support to sensible targets protection;
- Defense against chemical, bacteriological, radiologic, and nuclear (CBRN) attack;
- Reconstruction in case of anthropic or natural catastrophic events.

2.3 Network-Enabled Capability

In this landscape it is possible to frame the transformation of the military organizations many western nations have started since the early years of the new century. This transformation is involving both the national defense forces (including the ones projected towards outside areas of operations) and other governmental organizations, connecting together military and nonmilitary crisis units (foreign affairs, coast guard, border police, homeland protection forces, and so forth). Many of the activities of national defense, involving different government entities, are in progress. Following the lesson learned on September 11, 2001, when the majority of information was available but not intelligently synthesized, the development of an information infrastructure is underway in many nations. This new structure is aimed to link together all the military and civilian entities that contribute to national defense, creating a net-centric infrastructure, generally modeled and centered on the Defense Information Infrastructure (DII) for the following reasons:

- Use of international standards, like STANAGs (standardization agreements), ADatP (allied data procedure), NAF (NATO architectural framework);
- Indications and tools developed by the Network Centric Operation Industrial Consortium (NCOIC), in order to be interoperable with similar structure belonging to NATO and/or European nations.

To give a realistic idea of the state of cooperation among the most advanced industries of most advanced nations not necessarily belonging to NATO, tools already defined by NCOIC are reported:

- NCOIC Interoperability Framework (NIF): NIF provides a repository for organizing the enabling guidance developed by NCOIC. NIF affords interoperability implementation guidance to allow firms to develop system elements/nodes interoperable with systems/nodes being developed by others.
- Network-Centric Assessment Tool (NCAT): The purpose of NCAT is to measure network centric interoperability of reference models, architectures, systems, programs, organizations, or capabilities.
- Systems, Capabilities, Operations, Programs, and Enterprises (SCOPE): The SCOPE model is designed to characterize interoperability-relevant aspects or capabilities of a system or set of systems over a network
- NCOIC patterns: A pattern is a canonical solution to a problem. NCOIC patterns typically show relationships among entities involved in network-centric operations.

The process of transformation underway in many nations has the goal of reaching performances more suitable to the new international scenario for the fight against present and future threats. This transformation requires the focus on information superiority, rather than on heavy weapons. Indeed war and economy, while operating in different environments and with different purposes, have almost always used similar technologies to acquire information about possible targets, activities of enemies, and commercial competitors. The armed forces have always reflected the culture and the structure of the societies they belong to. In any postindustrialized society, information has become the strategic resource for excellence, such as heavy industrial assets, capital, and labor were in the first half of the twentieth century. Achieving information superiority through an information technology infrastructure for the acquisition and local distribution of the information is considered an indispensable capability of the armed forces to pursue an acceptable level of national security rather than relying on heavy weapons and number of soldiers. At the heart of this transformation there is the transition from single platforms of telecommunication and services to a communications network infrastructure, global and seamless, for the collection of information where available, and its distribution where needed in a secure environment. Without this infrastructure, the military instrument cannot operate in the new post–Cold War scenario, where it is not possible to predict how and where the armed forces could be used, because the threat may not always be associated to a specific geographical location or cyberspace. The terrorists and cyber threats, which represent the new enemies, can be anywhere in the world (even within national borders) and could attack either virtually or physically vital infrastructures of the nation like energy and water distribution, communications, transport, food chain, and so forth. This means that traditional areas of competence between military forces and homeland security forces are overlapping and, in accordance with the institutional roles of the various components, depending on the type of emergency, operations are lead by military organizations or civilian ones in close cooperation.

Following this change, the future armed platforms (air, land, and sea) will be lighter because they will not necessarily bring all the sensors they could need for a mission, but they will be able to receive and share information with other operational platforms through the global network, which they are connected to. A similar model can be used also for fire brigades, police forces, coast guard, and so forth.

The term platform is generically used to identify the combat or support units. These can be command and control, logistic and administrative systems, war and support ships, combat and transport aircraft, armoured and not armoured vehicles, battalions, brigades, platoons, single soldiers and in brief term Command, Control, Communications, Computers, Intelligence, Surveillance, Target Acquisition, and Reconnaissance (C4ISTAR) components. These components are permanently connected and identified and localized within the global network. Depending on their geographical location and combat/logistic capabilities, they receive their operational task and/or mission assigned acquiring the right to have access to relevant data from other sensors. Therefore, they become recipients of services and carry out themselves the functions of service providers for other platforms with their sensors. In this way the platforms share data in order to contribute to the visualization of an operational situation and share the operational awareness with the command and control centers of operations that can be joint national or international, fix or mobile, in country or outside. Each platform can therefore actively participate in the conduct of the operations.

2.4 Net-Centric Warfare Doctrine and Technological Challenges

In the military sphere, the integration of sensors and the sharing of information inside the communications networks are defined by the net-centric doctrine, and, as already mentioned, in general is not limited and restricted to national forces or borders, but extended if the threat is common to a cluster of nations or is transnational. The concept of network-centric warfare (NCW) was established within the United States in the mid-1990s (before September 11, 2001) as a central component of the 2020 Joint Vision strategy. From the platform-centric approach this doctrine leads to the new network-centric view, in order to acquire the information superiority in military operations, as it is already underway in the commercial world of the so-called information age.

The aim of network-centric warfare is to transform an information advantage into an operational advantage, exploiting the networking of systems technologically at the state of the art in order to enable different operational entities to interact according to specific procedures, using the following advantage's factors:

- The network structure improves information sharing;
- Information sharing increases the information's quality and, therefore, the shared consciousness of the situation (the shared situational awareness);
- Shared situational awareness in place enables collaboration and synchronization between operational entities, enhancing sustainability of action and speed of command (the operational tempo);
- Overall, a huge increase in the efficiency of the forces and effectiveness of the operations is reached.

The concept of net-centric warfare (i.e., fast transfer of information from sensors to effectors) from the initial function of mere support to operations, evolved towards effects-based operations (EBO), and has started to expand in a pervasive way, starting from command and control systems to the intelligence, logistics, medical, and administrative ones, according to the principle that everything must be available online and accessible. The expansion of possible applications, however, leads to prospects of a huge increase in information's traffic on the network (i.e., problems with bandwidth), with a dramatic cost increase; this leads to excessive amounts of information with the possibility to confuse the end users if the information is not intelligently filtered or managed. In order to operate in coalition with the United States (the country of reference in which the NCW concept was born) the other NATO and Partner for Peace (PfP) countries must be compatible with the U.S. net-centric system; however, it is for them economically unthinkable to reach and keep the same technological level. Thus, from the concept of net-centric, reduced models were derived, such as the network-enabled capability (NEC) of the United Kingdom, and NATO network-enabled capability (NNEC), the latter described in the NATO Feasibility Study (prepared in October 2005) and chosen by other nations, including Italy.

In general, the term network-centric warfare identifies a combination of doctrinal, procedural, organizational, technical, and human factors, which when appropriately integrated (or networked) obtain a situation of decision superiority on the opponent. The immediate consequence of this concept is to overcome the traditional way of conducting operations focused primarily on the use of platforms in favor of the use of systems. The aim is to create a computer network that can capture and manage in real-time databases and information coming from all available sensors and systems, for the benefit of decision makers and effectors physically dispersed, but interconnected.

The network infrastructure must have a capacity to process and store sensitive data not geographically concentrated to make the whole system less vulnerable to attack. The power of this network—this is the heart of the doctrine—is the synergy of its various components, as a direct result of access and information sharing. In the operational use, this capacity extends up to the acquisition of the shared situation awareness, which represents the most important success factor, especially in joint and combined operations. In relation to the three classical domains in which military operations take place (physical, informational, and cognitive), the establishment of NCW theoretically acts more effectively because:

- In the physical domains the various components are interconnected.
- In the information domains data are collected, synthesized, distributed, and shared.
- In the cognitive domains all have the opportunity to develop and share a better understanding of the situation.

Therefore, a force structured in a net-centric way, can operate in a synchronized manner over large geographical areas, with resources optimally sized and distributed in space, improving efficiency and effectiveness of its action, and enhancing the likelihood of success. In other words, although innovative, the idea of NCW has not supplanted the traditional criteria that underpin the conduct of military operations (i.e., the principle of unity of command), but allows the obtaining of some element that can be decisive for success in the conflicts of the information age.

Establishment of an IT infrastructure that supports the collection and distribution of information in real time, as required by the net-centric doctrine, requires the development of new technologies. Regarding communication systems, it is necessary to increase, in both the wired and the wireless (including satellite) domains, the available bandwidth as wide as possible, in order to transport a large amount of information on time. Moreover, the convergence towards a single communication protocol, IPv6 Internet Protocol, provides the appropriate priority and address to the information for its proper distribution. Everything must be done in an environment that ensures the security of classified information at the proper level of the information's quality. New technological frontiers are being crossed also in cryptography and it is not far until the day when quantic cryptography will become routine.

All current systems deal with individual data and are aggregated to provide the information requested. The future systems will be able to treat the information in order to constitute the knowledge base necessary for users who will need it. Nations are developing semantic engines to enable search and synthesis of relevant information. There are centers that are able to correlate seemingly unrelated data to provide solutions, or systems that identify, in fast streams of data, interesting messages in different languages.

The strategic role to be played in future IT infrastructure by the military makes it essential to develop technologies that help to reduce the vulnerability of such structures from attack (natural disasters, terrorist attacks, and so forth). All technologically advanced countries, which rely increasingly on information handling systems, have since taken steps that go under the name cyber defense, aimed at defining those measures essential for the use of IT systems and able to reduce their vulnerability.

Related to the previously mentioned technologies, cloud-computing nowadays provides a new supplement, consumption, and delivery model for IT services based on the network, and it typically involves the provision of dynamically scalable, and often virtualized resources as a service over the net. Cloud-computing is net-based computing, whereby shared resources, software, and information are provided to computers and other devices on-demand, like the electricity grid. The term cloud is used as a metaphor for the net, based on the cloud drawing used in the past to represent the telephone network, and later to depict the Internet in computer network diagrams as an abstraction of the underlying infrastructure it represents. Typical cloud computing delivers common business applications online; these applications are accessed from another Web service or software like a Web browser. Most cloud computing infrastructure consists of reliable services delivered through data centers and built on servers. Clouds often appear as single points of access for all consumers' computing needs.

2.5 The NEC Implementation in an Archetypal Nation: Italy

What has been so far presented could give the impression of a purely philosophical speculation or of a time-consuming process developing a complex system of systems that will be in use in a far away future.

That is not the case, because the NEC transformation is already in an advanced stage starting from legacy assets and looking at the future in an evolutionary approach: in other terms armed forces of western nations are evolving while operating in an open-ended process. Therefore, the following presents the situation in a real case well known to the writer: Italy.

Like other high-tech armed forces, the IT infrastructure of the Italian Defense is now composed of systems not always fully interoperable with each other, and in the absence of a single protocol, it is using an appropriate gateway for the exchange of information. Thanks to this, the Italian Armed Forces are already able to achieve a limited net-centric capability in specific geographic areas where special events (the G8 meetings, signing of important international treaties, and so forth) can be timely planned or in case of major natural disasters. However, the ultimate goal is to have this capacity full time for every situation and in any geographical location and to build a military C4ISTAR architecture, under the guidance of the joint chiefs of staff, to realize a seamless Defense Information Infrastructure (DII) able to interoperate with other similar international and/or domestic architecture, both military and civilians. Civilian and military personnel at different levels will interact by sharing information, situation awareness, and responsibilities.

This architecture should be flexible and should comprise different layers of interconnection to enable connectivity anywhere, anytime, if and when required. It will consist of building blocks of hardware, software, and services made available to different operational communities constituted for specific operations. At the end it will be a service-oriented architecture that can be modified to meet the needs of users.

The development of this architecture requires a considerable amount of time, based on national ambitions and available funding. The Italian defense has since decided that the building blocks of this architecture are the standards that NATO is setting, in order to ensure interoperability with NATO and other allied nations.

The modernization program of armed forces' C4ISTAR architecture is composed by several elements. The communication infrastructure, composed by wired, wireless, and satellite connections, is evolving towards a seamless environment Defense Information Infrastructure (DII); the three Command, Control, Communication and Intelligence(C4I) systems of individual armed forces are consolidating into a single joint C4I system that will support the Joint Operational Command (in Italian, COI, an acronym for Comando Operativo Interforze) as the highest operational level responsible for the conduct of joint operations. The interconnection of the three systems of the armed forces, the SIACCON for the army, the MCCIS for the navy, and the SICCAM for the air force, will be a system of systems, interoperable with their counterparts from NATO and other allied nations.

In regards to the army, the SICCONA, the command and control system mounted on board of armored vehicles, will be integrated with the SIACCON to take advantage of the services made available through the new software called Minor Level of Operation (MLO), which will allow in the near future the digitization of the battlefield up to the individual soldier used in operations. Indeed, a key element of the battlefield digitization is the Future Soldier program, which aims to provide individual soldiers with sensors, displays, and a system for exchanging data within the communication network.

Accidents caused by friendly fire (fratricide) are reduced through the implementation of two programs:

1. The New Generation Identification Friend or Foe (NGIFF), which operates within NATO, for the identification of friendly or enemy aircraft, and within the International Civil Aviation Organization (ICAO) for air traffic control;
2. The Battlefield Target Identification Device (BTID), which operates within NATO for the identification of ground vehicles.

In the future, there will be also systems for air to surface identification, currently under study and experimentation.

As far as the wireless component of the DII, The Italian Armed Forces rely on the development of software-defined radio (SDR) to achieve an effective wireless link, whereas the SDR is seen as the wireless pillar of future net-centric applications. The SDR will also be used in a special communications satellite. Indeed, satellite communications systems with SDR, rather than being limited to what is available at the design stage, can be easily updated from the ground during their lifetime while in orbit.

The satellite component has an important role in the transformation of communications: the Italian defense today uses satellite SICRAL 1, which has been in orbit since 2001, and the new SICRAL 1B, launched in 2009, which will work in tandem with its predecessor to deliver communication services also to the Carabinieri (Italian military forces police) on the Italian territory. Italy and France are now building SICRAL 2 that will replace SICRAL 1 when it is launched in 2012. Some communication skills present in SICRAL 1 have been moved from SICRAL 2 to ATHENA-FIDUS, which is another joint French-Italian satellite program scheduled to be launched in 2012.

From a service based on a single satellite, the Italian defense is moving towards a constellation of satellites that are complementary to ensure continuity of services both to national armed forces and governmental entities and NATO nations, Italy being in the consortium of nations providing satellite connectivity to the alliance. The satellite connectivity is used to let the Italian defense acquire advanced capabilities of telemedicine, operating in high definition and linking theaters with the medical resources and knowledge available at home.

The DII will share parts in common with the infrastructure of civil protection forces, known as Homeland Security Information Infrastructure (HLSII). Although the civil protection forces are independent of the armed forces, the exchange of information is being consolidated with the military forces. This includes sharing of existing databases and new data coming from observation satellites (SAR and optical) and reconnaissance crafts (manned and unmanned).

All this information is included in a common repository accessible to all authorized users, known as the territory's control system (SICOT), which will be able to provide information ranging from marine pollution data for the coast guard, the land pollution and financial crimes' data for the police, and the situational awareness for the military when and if needed. Both the armed forces and the civil protection forces will therefore benefit from a common framework of shared information.

For example, currently, the border patrol and coast guard have traditional sensors stove piped and the exchange of information is not automatic and depends on the degree and desire of cooperation among individuals. The future net-centric architecture enables both forces to enter a common repository consolidating, with appropriate institutional protocols, mutual cooperation: the information will be produced by someone, but it will be shared with everyone when and if needed.

The development of sensors for monitoring the territory not only improves military operations, but also helps to prevent, as far as possible, or mitigate effect of disasters such as earthquakes, volcanic eruptions, floods, and other natural disasters. Geologists in the future may be able to monitor forewarning signs like slight changes and tensions in the Earth's surface from tectonic movements or to solid tides, due to variations of gravity induced by the revolution of the moon, and in some cases have possible warnings of imminent catastrophic events. The industry in general, defense in particular, should play an important role in this transformation effort to build the future IT infrastructure.

The collaboration between the government and defense industry today is strong, but should be further increased. The future of complex systems requires a new frontier of systems in which new technologies are developed in a highly collaborative environment. In this new engineering approach, extensive use of modeling, simulation, and test bed is present, to assess, together with the user, the possible solutions prior to their development and acquisition. Only after a steady exchange of information between government workers, experts in operations, and the industry on various solutions, it will be possible to proceed to its full development, following the concept of spiral with successive improvements. In this framework, Italy is developing a Center of Excellence in Modeling and Simulation available for the other NATO nations.

2.6 Conclusions

The synergies arising from the close cooperation between government and industry must be international in nature through strategic alliances with European nations in order to meet new global challenges. Moreover, international cooperation is becoming even more essential, given the worldwide financial crises creating the new economic reality. Economic resources are declining today in any nation, as the recent G8/20 held in Canada has taught. In the future, further cuts may be enforced in research and development, acquisition, and operation. This places a greater emphasis on innovative synergies between governments and industry to produce at lower costs and higher quality.

Today the cooperation among the armed forces of different nations and the cooperation between the armed forces and civilian forces of the same nation for homeland security are mainly achieved by putting together representatives of the different organizations in the same operational room to coordinate field cooperation. In the future, following the NEC transformation, the operational systems of the different nations and organizations (military and civilian) will be federated through the network for common shared situation awareness. This new approach increases the effectiveness and the efficiency of the operations.

The challenge of NEC transformation is to put together all the building blocks in a coherent network-centric interoperable environment. Western nations' armed forces, Italy among them, have already done much to transform, but much remains to deliver efficient and effective systems. This transformation requires the focus on information superiority, rather than on heavy weapons.

In conclusion, the author may state that this chapter is essentially based on his experience and vision in the time he was the general director of the Italian Defence Communications (TELEDIFE).

Selected Bibliography

The information utilized in the frame of this chapter has been extracted from the classified and unclassified NATO circuit, subject to controlled distribution, and not available on the public circuit. Only information that could be disclosed without encountering any objections from NATO and Italian National Defense has been used.

For completeness, a list of documents available only to professionals in possession of the necessary need to know is listed here.

NATO Nations FORCE-GOALs:

- E2780: Enable Communications Network;
- E 2781: NNEC Information Assurance Framework;
- E 2860: Information Systems Network Enabled;
- E3023: Command and Control—Mode 5 IFF Capability;
- E2861: Network Enabled Services;
- E3070: Network Enabled Interoperability.

To govern the transformation process, the defense general staff of the archetypal nation (Italy) has published the "framework" strategy, consisting of internal directives of the Italian defense. These documents are not disclosed, except to insiders whose knowledge helped the author to compose this chapter. The considered documents are listed here:

"Final Operational Requirement of DII," (DII ROD).

"The Governance for Development of Defence Systems C4ISTAR, for the Harmonization and Optimization of Operational Requirements, Technical Standards, Requirements for Interoperability and C4ISTAR Security Sector," (SMD-NEC-003).

"The Guidelines of Modelling & Simulation for the Development of C4ISTAR Defence Systems," (SMD-NEC-001).

"The Methodology for the Definition and Development of C4ISTAR Defence Systems Architectures," (SMD-NEC-002).

"The National Development Plan of C4ISTAR Components Defining the Roadmap for the Transformation and the Priority Areas of Intervention," (SMD-NEC-004).

"The Vision of the Net-Centric Transformation."

3

Homeland Security and Challenges in Information Systems

Bernhard M. Hämmerli

3.1 ICT and Society, ICT and Control Systems, and the Blurring from Both

The development of critical infrastructures in the past 20 years has brought information and communication technologies (ICT) into nearly every component of critical infrastructure as well as into offices, homes, and user devices such as cars. ICT has given society efficiency and comfort on one hand, but added complexity and vulnerability on the other hand. As an example, in personal life, we are used to accessing ICT components every few minutes, such as cell phones, elevators, cars, Internet services (e.g., e-banking or just a simple topic-related search), washing machines, light regulators, and so forth. In public life we talk about critical infrastructure sectors such as electricity, transportation, finance, and water, which provide crucial services to the society as whole. A blackout of one or even several critical systems at once, each one enabled through ICT, is considered today as very unlikely and the imagination of a long lasting and multiple infrastructure blackout including all the consequences is not really dealt with by the general public. However, without any doubt such a scenario may happen, and national governments try to prepare for such situations. If preparation is studied, often simulation games are used to train the country's main infrastructure responsible for crisis preparedness.

3.1.1 Chapter Overview

Homeland security and challenges in information systems are discussed, initially depicting the landscape of information security and its interrelation to the critical infrastructure (CI) and critical information infrastructure (CII) or the combination of both C(I)I. When it comes to infrastructure protection, the according acronyms are CIP, CIIP, and C(I)IP (for definitions see [1]) (Section 3.1).

E-banking security and processes are used as an example to illustrate the information security challenges and its associated impact on society: we learn from e-banking about confidentiality and the meaning of identity theft and its countermeasure, which are all relevant for C(I)IP (Section 3.2).

With the analysis of the financial critical IT systems within Europe, as performed in the PARSIFAL project (http://www.parsifal-project.eu/), a broad European-wide discussed consensus on critical financial infrastructure challenges is presented. The eight identified fields that need research in the next five years are disclosed in the form of the recommendation (final report to the European Commission—information security research funding unit—www.parsifal-project.eu/images/PublicDeliverables/PARSIFAL_Summary_Final.pdf Page 3—(Section 3.3).

As a well-recognized fact, information sharing is a key activity for increasing the level of resilience, robustness, preparedness, and consequence management such as business continuity, and disaster and recovery planning. Key factors in building up information sharing centers and the trust within the sharing group will are discussed (Section 3.4).

One specific consideration on the balance of security measures and privacy concerns is given, just to prevent potential unlimited demand to control and monitor individuals (Section 3.5), before discussing overall conclusions (Section 3.6).

3.1.2 Critical Infrastructure and ICT

Intuitively everyone has an understanding of the terms critical and infrastructure as well as the combination of both. To sharpen and align the understanding the definition of the European Commission[1] is reproduced:

> Critical infrastructures consist of those physical and information technology facilities, networks, services and assets which, if disrupted or destroyed, would have a serious impact on the health, safety, secu-

1. Commission of the European Communities, "Green Paper on a European Programme for Critical Infrastructure Protection," Brussels, November 17, 2005, COM(2005), 576 final, p. 3, http://www.libertysecurity.org/IMG/pdf/EC_-_Green_Paper_on_CI_-_17.11.2005.pdf.

rity or economic well-being of citizens or the effective functioning of governments in the Member States. Critical infrastructures extend across many sectors of the economy and key government services.

Typically, the critical infrastructure (CI) is divided into 5 to 20 sectors, depending on organizing entity of the member states (United States and European Union). Typical sectors are according the EU definition (http://europa.eu/legislation_summaries/justice_freedom_security/fight_against_terrorism/l33259_en.htm):

- Energy installations and networks;
- Communications and information technology;
- Finance (banking, securities, and investment);
- Health care;
- Food;
- Water (dams, storage, treatment, and networks);
- Transport (airports, ports, intermodal facilities, railway and mass transit networks, traffic control systems);
- Production, storage, and transport of dangerous goods (e.g., chemical, biological, radiological, and nuclear materials);
- Government (e.g., critical services, facilities, information networks, assets, and key national sites and monuments).

The sector concept helps to build groups that share a common crises language and understanding, having similar educations and are responsible as a part of their jobs to run a CI (often privately owned). Common language and common values are important to establish well-functioning collaboration. In spite of being competitors in the core business within a given sector, experts start after some time to understand that the security dimension of the critical sector is not core business and may be understood as a domain of mutual collaboration and support. Having reached this level, the work within the CI sector can start.

Usually first tasks to start with are the following:

- Making an inventory of critical infrastructure.
- Making an inventory of the critical services that must run, such that the CI can work (dependency analysis).
- Making a vulnerability and threat analysis: the conclusion is then the risks, when taking into account the probability of specific incidents.

- Defining a minimum level of service that should in any case be maintained.
- Defining countermeasures to improve resilience of the infrastructure on one hand and to add as a plan B disaster and recovery planning.
- Tests of the countermeasures with simulated scenario and exercises, typically with one given sector at the start, in a more advanced stadium between sectors.

After understanding the basic concept of critical infrastructure protection (CIP), the information security component of CIP, the critical information infrastructure protection (CIIP) can be introduced. Scanning again through the list of sectors above specifically with ICT dimension in mind, one can start to understand the dimension of the task. The EU Commission defines CIIP[2] as follows:

> The programs and activities of infrastructure owners, operators, manufacturers, users, and regulatory authorities which aim at keeping the performance of CII in case of failures, attacks or accidents above a defined minimum level of services and aim at minimizing the recovery and damage. CIIP should therefore be viewed as a cross-sector phenomenon rather than being limited to specific sectors. CIIP should be closely coordinated with CIP from a holistic perspective.

CIIP are "critical infrastructures [in] themselves or that are essential for the operation of critical infrastructures (telecommunications, computers/software, Internet, satellites, etc.)."[3]

To distinguish between CIIP and corporate information security, we need to understand information security in more detail. The goal of corporate information security is to maintain the following properties within information systems:

Basic properties:

- Availability;
- Confidentiality;
- Integrity.

2. "Green Paper on a European Programme for Critical Infrastructure Protection," Brussels, November 17, 2005, COM(2005), 576 final, p. 19, http://www.libertysecurity.org/IMG/pdf/EC_-_Green_Paper_on_CI_-_17.11.2005.pdf.

3. Commission communication, 2005, p. 19.

Composite properties:

- Nonrepudiation (traceability, monitoring, logs);
- Authentication (who is acting);
- Compliance (local regional and global laws and standards).

Corporations want to minimize their economical losses caused by information security incidents. It is not about avoidance in any case of incidents, but it is about awareness of ICT risks and to guard risks and—if necessary—to increase countermeasures. The background of corporate information security relates always to maximization of revenue considering all aspects including investments into corporate information security, loss caused by information security incidents, and reputation. However, societal impacts caused by a blackout of CI sectors are not subsummarized within the umbrella of corporate information security. The investment of the delta when increasing from corporate information security to CIIP level has, of course, associated cost. Those costs are part of the debate—who should carry those costs? Some options are given:

- Regulation: Laws make the CIIP provider and operator carry the costs.
- PPP: Public–private partnership model (discussed later into more detail) operates on partnerships, where each contributor carries its own costs.
- Governments pay for the additional costs.

Historically, up to 1990, most critical infrastructure was state-owned and was provided by one single corporation. Governments were labeled as too expensive and being not sufficiently efficient. At this time, competition was considered to be the solution to increase efficiency and to reach lower prices. As we know, both situations eventually happened, but with a subsequent loss of control on the critical infrastructure. After many years of deregulation of critical infrastructures, national governments started to understand the challenge to provide an additional CIIP level of security to the corporate security of multiple companies in a deregulated market.

The attempted goal is to act on both, increasing information security in general as well to increase resilience and preparedness of CI and CII.

3.1.3 Critical Infrastructure and Control Systems

In recent years, operators of any infrastructure have started to connect their control system at least to the internal Ethernet, which is in most cases somehow connected to the Internet, or even directly to the Internet. The basic background for this connection is to save costs by:

- Using the internal net instead of adding a second and separate cabling;

- Taking advantage of networked infrastructure by inspecting the control system data through applications and user interfaces anywhere in the world. However, the risks associated with this connection type are in most cases not already assigned properly, and mitigation of risk is often not, or only poorly, addressed.
  - *Example 1:* Direct dependency electricity production and distribution. For metering and control system maintenance an Internet connection has been installed at many places during the last few years. The risks associated with this connection are not sufficiently assessed in many cases. For example, in the Idaho National Laboratory it was proven that a large diesel generator can be completely destroyed by Internet attacks. Other experiments executed by the Joint Research Centre Ispra of the European Commission have proven that pretty much all known vulnerabilities (DoS, Trojan horse, viruses) can be exploited in metering and control system applications with the utmost large effects on the given infrastructure [2].
  - *Example 2:* Indirect dependency. EU energy sectors' trading platform for electricity. This trading platform was based on the public Internet. Risk assessment disclosed that trading is essential to operate the electrical grid properly, and without trading, it is very difficult to control the flow in the transmission lines. This example demonstrates an Internet dependency of the energy sector.

Control systems are using real-time computing (RTC), or reactive computing, which have hardware and software real-time constraint—that is, operational deadlines (typically milliseconds or even less) from event to system response. By contrast, a nonreal-time system is one for which there is no deadline, even if fast response or high performance is desired or preferred. A real-time system may be one where its application can be considered (within context) to be mission critical. A real-time deadline must be met, regardless of any circumstances such as system load.

Supervisory control and data acquisition (SCADA) is the term often used in the context of homeland security. SCADA generally refers to an industrial control system: a computer system monitoring and controlling a process. The process can be industrial, infrastructure, or facility-based and Wikipedia (http://en.wikipedia.org/wiki/SCADA) defines the application field as follows:

- Industrial processes include those of manufacturing, production, power generation, fabrication, and refining, and may run in continuous, batch, repetitive, or discrete modes.

- Infrastructure processes may be public or private, and include water treatment and distribution, wastewater collection and treatment, oil and gas pipelines, electrical power transmission and distribution, wind farms, civil defense siren systems, and large communication systems.
- Facility processes occur both in public facilities and private ones, including buildings, airports, ships, and space stations. They monitor and control heating, ventilating, and air conditioning (HVAC) access and energy consumption.

SCADA systems have two properties which make them much more difficult to defend against Internet attacks than normal information systems:

- A continuous computational effort with well-defined deadlines ran on a hardware, which is minimal and cost optimized (i.e., there is not much computing power left for security).
- Because of the time criticality and always performing the job often for several years, the usual update mechanism such as those used in Windows systems are not admitted. The system may not be interrupted or restarted at any time it performs. Vulnerabilities cannot be fixed, even when well known and patches available.

Of course, there are some solutions and workarounds available by protecting the control entities with particular system architectures and front-end computers, which terminates links in an appropriate way. However, the community is still far away from optimal and off the shelf solutions.

3.1.4 The Mobility Trend

Today, nearly everybody is aware that the Internet is mobile. iPhones, iPads, and Internet newspapers are just the first indicators of this trend. Soon the number of mobile endpoints will be more than the fixed net endpoints. What is the influence on C(I)IP of the mobile trend? Some answers:

The hackers can use mobile endpoints to start attacks. Usually, this type of endpoint is less protected than the fixed net, and also has less computer power to protect itself. The update mechanisms are by far less advanced in mobile devices and used in much larger intervals, if at all.

The sheer mass of mobile endpoints and the fact that traceability of attackers is much more difficult give the attackers more protection. Cross-border mobile attacks are more difficult to investigate, and need more elapsed time.

The mobile infrastructure, cell phones, and Internet are so much part of the everyday life that a long-term operational stop would harm the society and its well functioning with high impact. Therefore, the mobile infrastructure will be soon a critical infrastructure as well.

The mobility trend and the according mobile communication infrastructure is a second communication infrastructure complementing the fixed net communication—an alternate communication channel (e.g., for backup purposes in some failure situations)—and a new infrastructure society depends on and uses it for critical application. For example, text messages are applied in Switzerland as a part of the communication between patient and doctor for life critical applications, and for certain dosing of medicine. However, the specification of the delivery of text message is "best effort" in Switzerland. There is no guarantee of a timely delivery or a guarantee if delivery at all. But in normal operation, text messages are very reliable and various sectors start to use text messages for critical applications.

In general, the meaning for C(I)IP of the mobile infrastructure is not perceived sufficiently. More work is urgently needed on this topic.

3.1.5 Blurring Borders of ICT, General, and Mobile Infrastructure

At the start, IT-systems were local and very much restricted to a location. With the advanced networking started in the mid-1990s, the IT systems were connected for services as e-mail, Internet, file serving, and specific corporate applications. Around 2000, mobile computing, first with laptops, and later on with handheld devices, started to spread out. In the meantime, SCADA systems were increasingly more connected to the corporate network and accessible over the Internet. A proof of this fact was the documentation of the Chinese penetration of the American power grid (http://online.wsj.com/article/SB123914805204099085.html). As stated in the source document, "Over the past several years, we have seen cyber attacks against critical infrastructures abroad, and many of our own infrastructures are as vulnerable as their foreign counterparts," Director of National Intelligence Dennis Blair recently told lawmakers. He stated additionally, "A number of nations, including Russia and China, can disrupt elements of the U.S. information infrastructure." This means that such attackers had reached a certain level of electronic ownership of the SCADA U.S. grid infrastructure that they would be able to shut specific parts down at times they wanted. This quote serves as an example of the blurring border of ICT in a global context. Furthermore, on the attackers side (example above), with appropriate applications, the shutdown command could be executed with a handheld mobile device, wherever this device is located in the world. With SCADA technology, it is also obvious that the borders between ICT and general infrastructure (here the electrical grid) are blurring in an ICT integrated sectorial infrastructure. In fact nearly any infrastructure today is ICT-optimized or ICT-supported, which finally results in a new view of critical infrastructure: the ICT integrated infrastructure. In the defense sector, network-centric warfare is recognized as the best way to gain better control. However, with a potential outage of the ICT infrastructure, not much is operational anymore—another fact to make the essential meaning of critical information infrastructure and its integration in other sectors very compelling. Any borders, inside and outside an IT infrastructure, may melt down and result in a more generalized view on ICT. ICT is today part of any sectorial infrastructure.

3.1.6 Globalization and Virtualization

As the example of the electrical grid demonstrated, the ICT dimension brings a global view on any infrastructure. However, more than on any infrastructure, the data center is globalized in such a way that it is hard to understand what infrastructure we use and where in the world it is located. Technologies such as virtualization allow seamless switchover of applications and a virtual server on a global-scale infrastructure. This means that the users do not know whether a specific job is running in India, the United States, or in Europe in a data center. However, for critical application this might not be what users or governments wish.

To reinforce this statement, the likes of Facebook and Web servers may be analyzed; we might use both daily but we do not know where these applications are running, neither on which platform nor which type of software is uses. We just receive the functionality we desire. This is proving of the appetite on functionality without assessing risk.

The trend of cheaper globalized services versus the wish to have control on a system and contents are competing in many organizations and in national governments. Moreover, economical pressure is not always well resisted, even when the backup side has to be evaluated and finally ends up somewhere in the world not knowing what this means in case of an outage.

It is of greatest importance to analyze system and contracts of systems in utmost detail to understand their architecture, geographic distribution, and the associated risks.

3.1.7 Need for Actions on the Policy Side: C(I)IP on the International and European Level

Over the past few years, the attention of policymakers and industry players towards the protection of critical infrastructure has grown remarkably.[4] In the

4. The task force CIP of the Centre of European Policy studies CEPS in Brussels dealt with these topics, and served as a source: www.ceps.eu and www.ceps.eu/taskforce/critical-infrastructure-protection-eu.

era of networks, citizens and businesses have become increasingly dependent on a large set of infrastructures encompassing energy networks, the banking sector, telecommunications, the Internet, and so forth. In the United States, a national program to protect critical infrastructure was launched in 1998 under the Clinton administration, and was confirmed and updated under the Bush administration in 2003. At the EU level, the European Council of June 2004 called on the European Commission to prepare an overall strategy to protect critical infrastructure. The Commission adopted on October 20, 2004, "a Communication on Critical Infrastructure Protection in the Fight Against Terrorism," which put forward suggestions on what would enhance European prevention, preparedness, and response to terrorist attacks involving CI.

The CI debate continued with the Council conclusions on "Prevention, Preparedness and Response to Terrorist Attacks and the EU Solidarity Programme on the Consequences of Terrorist Threats and Attacks" in December 2004, which endorsed the intention of the Commission to propose a European Programme for Critical Infrastructure Protection (EPCIP) and agreed to the setting up by the commission of a Critical Infrastructure Warning Information Network (CIWIN). After a green paper adopted in November 2005 and the December 2005 Justice and Home Affairs (JHA) Council Conclusions on Critical Infrastructure Protection, the Commission finally adopted a communications, which designated European critical infrastructure as those ones that, in case of fault, incident, or attack, could impact both the country where it is hosted and at least one other European member state (COM(2006) 786, http://eur-lex.europa.eu/LexUriServ/site/en/com/2006/com2006_0786en01.pdf).

The CI and CII definition of the EU Commission was kept deliberately broad to enable a horizontal, multisectoral policy approach. It is increasingly accepted that both CI and CII face potential threats; an attack on either would result in significant economic costs. For example, The World Economic Forum estimated in 2008 that there is a 10% to 20% probability of a major CII breakdown in the next 10 years, with a potential global economic cost of approximately $250 billion: is the financial turmoil such a breakdown? Similarly, the U.S. Business Roundtable in 2007 suggested that the economic costs of a month-long Internet disruption to the United States alone could be more than $200 billion, while the Organization for Economic Co-operation and Development (OECD) estimates the annual losses to U.S. businesses caused by malware to be $67.2 billion (EU Commission Presentation, Andrea Servida, 2007).

On June 15, 2010, the Senate of the United States has amended the Homeland Security Act of 2002 by Protecting Cyberspace as a National Asset, Act of 2010 (hsgac.senate.gov/public/index.cfm?FuseAction=Files.View&FileStore_id=4ee63497-ca5b-4a4b-9bba-04b7f4cb0123 and www.zdnet.com.au/internet-kill-switch-proposed-for-us-339303838.htm). The act includes education of the workforce by creating centers of excellence, building up and enhancing cyber forces in police and defense. This is a justification of the growing importance of ICT towards an essential critical infrastructure having impact on any other sector.

3.2 The Information Security Challenge: E-Banking as an Example

Over the past few years, there has been an arms race between attackers of e-banking systems and the corporate defense measures. As the term arms race implies, it is very difficult for both sides to win with today's usual technology. However, with large additional investments, it is feasible to secure e-banking in the future against qualified malicious attacks.

The first sad fact is the asymmetry of the arms race when it comes down to the game between attacker and defender. The attacker is able to use the weakest link or the last hole in the security concept; meanwhile, the defender has to use perfection on all IT systems and drive the security process from personnel to operations processes and finally ending up in securing systems. Especially in respect to securing systems consisting of millions of lines of code, the old rule that errors offering attack surface repeats every 50–300 source code lines is still true. This fact makes the insiders humble and aware that systems may be secured, but remain somehow vulnerable, even with strong countermeasures.

E-banking is a very well researched field and serves in this context as an example to demonstrate the arms race, the countermeasures, and around 15 years of strong effort to secure the system, which may be—first in European states—completed and true in this decade. The challenge to design just a—more or less trivial—browser application that resists all attempts of manipulation was extremely challenging. From this example, the reader may imagine what it means to secure more complex systems. As already indicated, the e-banking system resists with the newest technology only manipulation, but not qualified attempts to intercept the information flow.

First, we will explain why and what makes the securing of e-banking so difficult. Then, countermeasures are discussed and conclusions finish this section.

3.2.1 The Challenge to Secure General-Purpose Computers for E-Banking

Today, clients' e-banking infrastructure is a general purpose personal computer with an Internet browser. The design of this system allows both secure and insecure channels to the Internet. Therefore, we have systems that are used for confidential secured applications, but also for general purpose computing, like office applications, e-mail, and possibly for telephony applications, video streams, and so forth [Figure 3.1(a)]. Today, the two attack methods that have

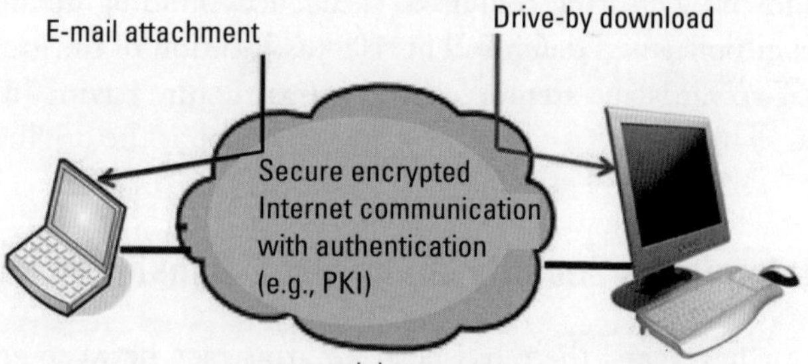

(a)

(b)

Figure 3.1 (a) Combination of secured black channel and insecure red channel used for attack. (b) Symbolic representation of secure systems with insecure side channels allowing circumventing of the security measures.

no waterproof countermeasures are e-mail attachments and drive-by downloads. An e-mail attachment, today for secure purposes mostly a PDF, may still have an implanted Trojan horse, which allows the attacker to get, via the insecure firewall and virus tested channel, remote access to the client PC and enables manipulation. Drive-by downloads introduce also Trojan horses via the usage of a Web page that has implanted malicious code to infect the PC with a Trojan horse. Both attack technologies uses malware, which is not recognized by the internet security software such as firewall and virus or malware control. Figure 3.1(b) demonstrates this fact symbolically. The black market of malware offers such attack software at a very low price, including a license agreement. This malware is used as an implant into PDF or in a Web page of an insufficiently protected Web server. As a result of these efforts, the attacker has gained system access to a victim's computer at relatively cheap price.

In the next step, the attacker creates a copy of the login Web page of the clients' bank. This copy has to look exactly identical as the original Web page, such that the client still believes he is connected with the bank. Therefore, the user trusts the bank and provides his credentials for gaining access. However, instead, the credentials are provided to the attackers instead of the bank on the other end of the Web page. As a response to the received credentials, the client receives a message "Web site temporary not available, please try again later." As every Web user knows, that services may be once upon-a-time out of service, normal clients accept this message and carry out commands as recommended. Now the attacker, who has committed identity theft, can access the bank with the clients' credentials and transfer the money to an account for his own profit.

The Web server of the attackers is usually not their own server. Normally, they use any other server, which is not sufficiently protected to display to the client the banking Web site. Such servers might be used also as a drop zone, where the attacker stores his credential and picks this up with a much lower probably of being caught.

What we learn from this example:

1. The combination of a secure and insecure system results always in the lower trust level, in this case the insecure.
2. The best identity management/authentication methods, encryption, and other measures do not help when the attacker owns authority on the clients' PC. Attackers circumvent the security measures.
3. User awareness and education does not help either.
4. The attacker operates via intermediary systems called drop zones, which cannot be efficiently stopped or traced in the midterm. Moreover, it seems to become a fact of life.
5. Attacks are relatively cheap to perform.
6. Only a minority of servers could have been shutdown with the support of the ISPs and/or local law enforcement agencies.
7. For certain jurisdiction it is deemed impossible to get the attackers arrested even though their identity might be revealed.

The new countermeasures for this situation will give some hope.

3.2.2 Countermeasures Against E-Banking Fraud

As we have seen in the section, the usual security concepts are of limited value in the given e-banking situation; new ways to face the challenge have to be found.

There are basically three concepts, which are observed in industry, to act on the postincident side:

1. *Denial of money transfer.* Whenever attackers want to use their power, they need an account to transfer the stolen money to. Counterfighting this transfer systematically by rules and through extensive monitoring is a very successful practice. Attackers earn (normal attackers' business) and are discouraged by income cut after completing an otherwise successful robbery. Thus, the destruction of the business model of the attackers leads to redirection for their efforts to places where gaining criminal money is easier. The trick is to increase attackers' investment per gained Euro by fraud such that e-crime is not beneficial and the attackers are discouraged.
2. *Repression.* The exchange of criminal money transfer information results in early warning for the sharing entities. With the knowledge of this information, qualified regulatory pressure. Furthermore, if according laws and regulation are given in the state where the criminals act, successful requests for criminals' prosecution can be placed. The EU Cybercrime Convention would enable such scenarios within the European Union and this would lead to much higher pressure to the e-criminals. However, the information-sharing community is still fighting for final ratification of the cyber crime convention).
3. *Counterreaction by attacking the criminal's server.* Counterreaction works, but it is not really beneficial. It is too much effort to identify new behavior of the attackers and to reach a high observation level again.

Up to now, a combination of (1) and (2) was used and has led to a high degree of success. There also concepts in the preloss countermeasure domain, which lead to success. The basic idea of those concepts is to use any form of verification on a second channel—not accessible by the hackers—for each single money transfer. There are basically two ways of using verification on a second channel well-known today:

1. *Mobile TAN:* Mobile Transaction Authentication Numbers (MTAN) and other forms enabling the verification of identify on a separate channel, such as special hardware devices, are already able today to discover the above described fraud. Many of the major Swiss banks have testing underway of such verification methods, leaving choices for biometric, MTAN, or federated identity solutions with an independent second verification method. The down side of this method is that high volume transactions are very difficult to handle because large quantities of verifications must be provided.

2. *Account accreditation:* The idea is to avoid the payment to the hackers account. Each account has a list of admitted accounts as beneficiary. This method has the advantage that the numbers of verifications decrease fast because many transactions repeat. Also a list of preverified accounts such as electrical power, telephone, rent of apartment, and so forth, may be generated from history. The verification can be made (e.g., with a challenge response procedure) where the response is calculated in an offline device.

The good news on e-banking is that the security is permanently increasing, such that manipulation may be excluded finally. However, confidentiality is not guaranteed, as long as it runs on a general purpose system colocated with insecure application and network access.

Reflecting on homeland security, e-banking serves as an example of how highly secured and mission critical systems are vulnerable. Furthermore, our society has changed and mass applications such as e-banking are in many member states an essential critical infrastructure that cannot be turned off for a month. Paper-based banking and its infrastructure is often reduced by more than 80%. The actual capacity that would be needed to switch back to paper flow is not available anymore and cannot be rebuilt in a reasonable timeframe. A blackout for a longer period of e-banking, as well as other Web-based applications, would downgrade our society in way that the economy and individual would really suffer, resulting in demonstration and uncontrolled political movements.

3.3 Analysis of Research Needs in Critical Financial Infrastructure: Parsifal

Critical infrastructure and services in financial industry are important for our society and the financial industry starts to understand the topic beyond the normal and well maintained business continuity management and disaster recovery plans (BCM&DRP). In 2008, the European Commission funded the Coordination Action project Protection and Trust in Financial Infrastructure[5] (Parsifal; www.parsifal-project.eu), whose purpose was the composition of a strategic research agenda in the critical financial infrastructure (CFI) domain. Within Parsifal, through two workshops and verification by a stakeholder group of more than 150 financial and IT experts, key recommendations and findings were produced.

5. Parsifal project team members are indirect coauthors of this section and permitted kindly the use of projects documents to write this section. Thanks to Rafael Llarena and Aljosa Pasic (ATOS Origin Spain), Susan Morrow and Sandy Porter (Avoco Secure), Henning Arendt (Arent Business Consultant), Jean-Yves Gresser and Tom Buschman (Edge International BV), and James Clarke and Kieran Sullivan (Waterford Institute of Technology Ireland).

3.3.1 Project Overview

The financial sector is one of the leading sectors in secure distributed and worldwide secure mass applications; many other sectors can learn what is pending and how to improve from the financial sector. That is why it was determined that the analysis of the financial sector is quite important. Furthermore, criticality in the financial industry is not about one bank—which can always go out of business–when receiving national support because the bank is considered "too big to fail" (http://www.ceps.eu/ceps/download/3025; Overcoming "TOO-BIG-TO-FAIL"). The term criticality in the European context primarily interfaces between the financial services and how those are cooperating with each other is a topic to be researched.

To approach the Parsifal topics, the need to harmonize language between the business (financial) and the IT community was continually raised by the stakeholders. However, the language challenge is even larger: The various communities in financial institutions (e.g., business continuity, networking, and software development) use term differently. Furthermore, each country has a slightly different usage of terms. To realize common exercises across internal communities and boarders, a common language (i.e., terms and ontology) has to be developed. An introduction to terms (glossary) and how they relate to each other (ontology) is the first deliverable of the project and is on the project Web site available as any other deliverable ("Ontology of financial risks and dependencies within and outside the financial sector"; http://www.parsifal-project.eu/images/PublicDeliverables/parsifal%20d2.1%20draft%20ontology%20of%20financial%20risks%20and%20dependencies%20within%20and%20outside%20the%20financial%20sector.pdf) introduce the reader to the idea of using ontologies and glossaries to better understand the risks within the CFI and the key concepts.

Furthermore, a description of the current situation in EU financial services gave some insight into the operation of the financial sector, which is a prerequisite to start with more detailed work.

A first workshop was facilitated in which the research challenges of the financial sector were discussed with the experts in the following three topics-related groups:

1. *Controlling instant on demand business in CFI:* authentication, identity management, resilience, and denial of service;
2. *Entitlement management and securing content in the perimeterless financial environment*: identity, policy, privacy, and audit;
3. *Business continuity and control in an interconnected and interdependent service landscape:* compliance, protecting critical processes.

The discussions in the groups were twofold. First, the future scenarios were discussed, which need a change in security or more attention. The scenarios are the justification of why something could be more important in the future. Second, technologies to be developed, technology-related innovations, and research challenges were discussed. Finally, the mapping of the challenges to the scenario was elaborated, to sort out technology visions without any clear relation to improvements in the financial infrastructure. As a result, the graph mapped challenges in CFI protection to appropriate scenarios (see Figure 3.2).

The key ideas and paradigms in this diagram are:

- Infrastructure—technology platforms, increasing complexity;
- Infrastructure—business practices and working patterns, increasing complexity;
- Infrastructure—data security methodology, highly distributed networks, and borderless environments;
- Countermeasures—robust identity management (IDM), new threat/fraud recognition methods;
- Systems—identity mechanisms and identity management;
- Compliance—highly distributed networks and borderless environments, data security in highly distributed networks;
- Global threats versus local measures.

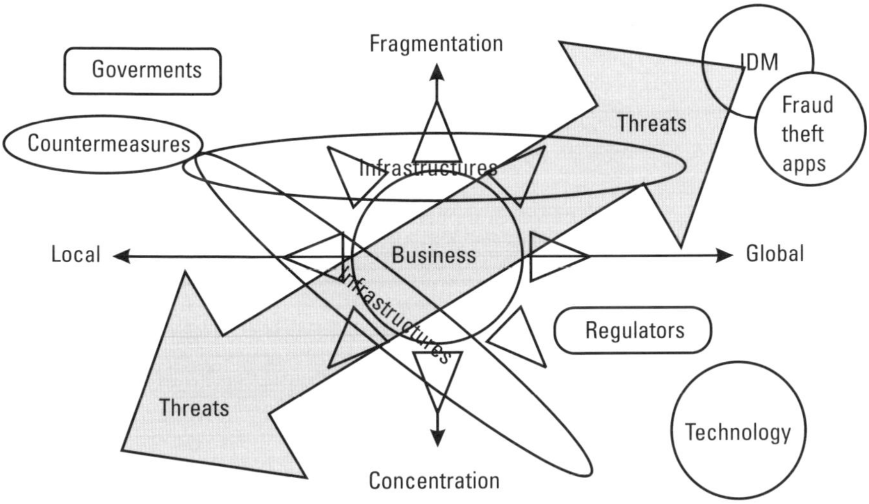

Figure 3.2 Condensed representation of a scenario and challenge that shows the main areas of concern, as directly expressed by the stakeholders.

The gained knowledge was condensed once more into eight principal recommendations to the European Commission for research investments, fostering the resilience and robustness of the European critical financial infrastructure (see Table 3.1).

Working with these recommendations, the questions arose regarding which are the most important ones, and how the dependency between the recommendations may be represented. The answer was the objective of a second workshop, as well as online consultation forms, where financial experts could

Table 3.1 Parsifal's Recommendations and Research Directions Associated to the Groups That Have Them Elaborated

Stream 1: Instant On Demand Business	1. Classification of identity attributes for online and mobile users of financial services should be defined and well understood by providers of these services and their customers.
	2. Trust indicators need to be developed, which allow for the various gradients of trust any entity might achieve when using specific financial services.
	3. Support platforms are needed for the management of multiple identities to allow consumers to authenticate themselves with various professional and private identity attributes.
Stream 2: Entitlement Management	4. Digital identities are required that are highly standardized across the financial services sector, with the introduction of mandatory IDs for all financial institutions, cross-border interoperability, and a "single/global" identity issuing authority.
	5. Data security measures are required, (1) such that a digital identity links directly with a security policy to a data object, (2) that data is secured as encapsulated entities, and (3) with flexible security policies that are based on individual access rights plus digital rights management (DRM) for enterprise content to allow for flexible security policies and geographic boundary control.
	6. New computing paradigms need to be analyzed, which allow for de-perimeterization of the organization (e.g., cloud computing), supported by any new security focus. Predictive models need to be created to understand security risks. Cross-border legal issues need to be resolved.
Stream 3: Business Continuity	7. Design and implementation of secure platforms and applications, which should include an alternative and secure communication system/infrastructure, to be overseen by adequate coordination response team(s) at a national and international level.
	8. Testing, design and implementation of such secure platforms, applications, and infrastructures through trustworthy exercises between CIP sectors and governments. Models for business continuity need to be extended to (1) sharing risks and (2) end-to-end communication between trade participants, as well as to (3) the volume and the complexity of specific financial markets. These models should be "crash" tested, regularly evaluated, and updated.

vote on the importance of each recommendation. The expert reconfirmed that all eight recommendations touch key points.[6] Figure 3.3 shows the time and content dependencies between each of the eight recommendations (i.e., how each recommendation/issue links to the other recommendations). These dependencies should be considered when deciding which area of research to fund and in which order funding should take place.

Further main documentations of the Parsifal project are:

- "Trusted Sharing of Confidential Information in CI" (www.parsifal-project.eu/images/PublicDeliverables/parsifal%20d2.3%20presentation%20on%20trusted%20sharing%20of%20confidential%20information%20in%20ci.pdf). It covers the most recent research in the area of securing data within CI and current developments in data security including digital identity.
- "Gap Analysis and Future Needs Report" (www.parsifal-project.eu). It defines the requirements for future research.

It was astonishing that four of the eight recommendations are associated with digital identities. The team did not expect this outcome, but obviously

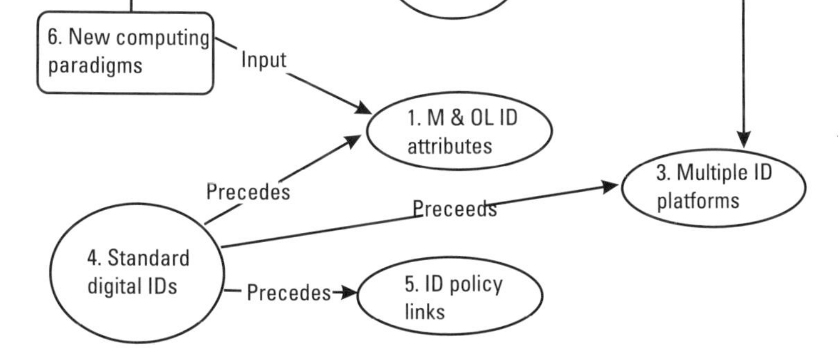

Figure 3.3 Dependencies between the eight recommendations.

6. See Figure 3 in www.parsifal-project.eu/images/documents/PARSIFAL_Marketing%20Document.pdf.

secure digital identities are found to be a key base functionality on which object-related security policies and entitlement may build on. Furthermore, simulation and exercises across borders and across companies must be continuously adapted to the newest technologies and needs ad hoc such as virtualization and cloud computing. Last, but not least, the end user has a major influence on the whole security discussion: informing an end user on risks to educate him to a risk-based behavior could improve the actual situation.

3.4 Introducing Information Sharing: Key Challenges

Information sharing is by today an internationally accepted concept to counterfight large scale attacks on infrastructure. The information that should be exchanged relates to the strategic and tactical security status of CIP sectors and is meant to be exchanged primarily in a closed group, consisting of main representatives of the service operators/providers and regulators, with some support from experts and suppliers. The group that exchanges information has a standardized name: Information Sharing and Analysis Center (ISAC) (www.isaccouncil.org). Normally, members are employees at an executive level for the most part, dealing with strategic information, risks, and vulnerabilities as well as forensic investigation and prosecution.

In ICT at the operational level, information exchange of actual threats, vulnerabilities, patches, updates, and so forth is done in computer security incident response teams (CSIRT) (www.cert.org/csirts/csirt_faq.html), the general and unprotected version of a computer emergency response team (CERT). The members of CSIRT are technical educated and highly skilled experts in online computer defense and attacks. Corporations and national entities are running CSIRT as well the military organizations in many countries for its own highly specialized infrastructure. CSIRT are well networked within Europe and the United States, and usually cooperate well with each other, as well as with the suppliers of ICT technology.

ISACs need a link to CSIRT to keep track of actual incidents, and also as the instrument to react to cyber incidents at the technical level. ISACs may react on other levels like organizing different alternatives to diminish the impact of specific incidents or trigger specialized organizations for disaster management. ISAC members have also a translation task to be intermediaries between subject matter experts and senior executives or governors.

The suppliers of hardware and software, as well as of security systems, play a key role in ISACs and CSIRTs. At the end, many reactions, fixes, and improvements must be done in collaboration with suppliers. Furthermore, it has to be taken into account that, in case of a worldwide disaster in cyberspace, suppliers must set priorities to whom they talk and whom they support first.

This is a fact that many nations and organizations are not sufficiently aware of today, and these entities could be surprised to have to wait for supplier support if an incident occurs.

The challenge of information sharing was approached by the ICT research community in Europe with building an according infrastructure to exchange, in various grades of trusted communities, confidential information. For instance the Joint Research Centre (JRC) of the European Commission in Ispra (http://ipsc.jrc.ec.europa.eu/activities.php?id=1, http://ipsc.jrc.ec.europa.eu/activities.php?id=2) has developed such software, as well as the European Commission cofunded ICT- FP7 CoMiFin (www.comifin.eu) project, which is still running and is exhibiting a similar approach. The CoMiFin and aforementioned Parsifal projects had very close links. However, the main obstacle to introduce information sharing is obviously not the exchange ICT platform: moreover, the real challenge is motivating the right community personalities and building willingness and trust between those personalities. For this purpose, the European Network Information Security Agency (ENISA), together with RAND Corporation, facilitates a workshop with the purpose to arrive at a robust list of prioritized incentives and challenges for information sharing between organizations in trusted groups (such as in information exchanges). This list and the evidence gathered at the workshop are expected to develop a set of recommendations and European policy priorities for further action (www.enisa.eu, http://www.enisa.europa.eu/act/res/workshops-1/2010/enisas-and-rand-europes-workshop-on-incentives-and-challenges-for-information-sharing-brussels-1st-july-2010).

Some countries such as the United States, United Kingdom, Switzerland, the Netherlands, Sweden, and Germany [1] have successfully started information sharing. In Switzerland, the author of this chapter was the initiator, lobbyist, and advisor when building up the information sharing center MELANI in the time frame of 1998 to 2005. From this experience, key success factors are discussed.

In our first argumentation, we did not argue with security needs as most of the initiatives do. We were arguing with economics of information sharing and increasing the prevention and reaction capacity, by allocating this capacity where it is needed in the case of an incident. The basic idea was that it is too expensive to facilitate in each of Switzerland's largest companies [many of those are in Forbes Fortune 500 (http://en.wikipedia.org/wiki/Fortune_Global_500, www.forbes.com/2010/04/21/global-2000-leading-world-business-global-2000-10_land.html) worldwide] a sufficiently well-staffed team. It would be much more beneficial to group the forces. The second point in argumentation was that attacks will come from outside Switzerland. The Department of Defense, which would be in charge of exterior security, had at these times no well elaborated role in cyber defense and corporations could not expect or rely on protection against incidents or support after an incident had happened. With these arguments, the government successfully started the MELANI project. Thus, it was a credible value proposition which caused the takeoff.

When information sharing is about to start, it is very important that very high-level experts are included. Champions are asked, which create a followership, personalities with which every security experts would be proud to collaborate. Such persons are motivation and credible sources of the trust that must be gained. The Areci (www.psc-europe.eu/index.php?id=librarypublic&filename=PSCE-RD-024_Annexes.pdf&dir=Reference_documents&task=download&mountpoint=6) study from Alcatel Lucent, written in 2007 for the European Commission, gave many recommendations on how to start information sharing groups, but after long discussion with the project leader, the acquisition of champions was identified as one of the very critical issues for success.

Mutual trust in information sharing is the key issue—building it up and maintaining it over time is a huge and ultimate challenge. Approaching the trust issue, it is important to realize that trust is at the end nothing else than a feeling and a personal attitude, which is as delicate as a young growing plant. The following properties describe it:

- "Trust is coming like a snail but going like horses." This wise phrase points out that building up trust is a long-lasting process in which cultural expectancy must be met and aura of faith, truthfulness, and acceptance has to be developed. In the case of breaking trust, trust will disappear fast and recovery is very difficult and challenging. These facts relate to the experienced-based component of trust, which is also called confidence.

- The other component of trust is rule based. Consensus on rules for monitoring and evaluating trust, having planed penalties when breaking trust, result in easier adaption to trustful exchange between members of the trusted circle.

- Any attempt to give a military order for trusting each other fails, is counterproductive, and results in silent opposition.

Minimizing potential vulnerabilities when trusting each other can be handled with two basic concepts:

- *Anonymization:* Change of names and institutions in the same context;

- *Sanitarization:* Change of names and institutions in a substituted context—Sanitariation does not allow any conclusion towards the original context. However, the described substitution context should result in equal consequences as the original context, when planning countermeasures and reactions

Anonymization and sanitarization enable describing situations and incidents in a way where trusting persons minimize their risks and exposure. Preparation to apply both concepts should be made.

The willingness of mutual support and sharing in crisis situations depend on the preprepared cooperation. Natural human behavior returns to egoism (you or I) the deeper the crisis situation is. This fact may be observed, for instance, in extreme mountaineering situations, such as performing above 8,000 meters of altitude. Every person is exclusively on their own and may not expect to get help from the others is in many teams an agreed rule. To overcome this natural behavior, responsibilities must be allocated before incidents, and an expert network must be set up and tied together prior to an incident. Confidence and trust must be elaborated before crisis, should be tested in crises, and lead to more trust and confidence when successfully applied.

Switzerland has a very good initial situation to generate trust, because many higher-level managers and experts are linked in various networks together (e.g., university, policy, military, and sports). Through these networks, and the given relatively small size of the country, trust is a natural element of the society, and develops generally faster than in other countries with larger size or more challenges in its past history.

The MELANI concept and operation were reviewed by ETHZ (a neutral body) twice, in 2007 and in 2010. It has been shown that only a single case or a few successful cases of information sharing avoid so much damage and that the prohibited potential damage paid off, for some years, the operational cost generated by running information sharing in the trusted circle of MELANI.

The success of information sharing in Switzerland is based on clear, credible, and realistic argumentation when presenting the initial concept, as well as of some nontransferable characteristics (to other countries) of Switzerland's culture and society.

3.5 The Dilemma of Security-Privacy

Introducing the topic, general positions of privacy and security sides are described. Out of the many facets of the given dilemma, in this section the asymmetry of the development of both privacy and security is discussed, and measures to compensate this asymmetry are proposed. In this book other chapters cover the overall topic in detail, in particular, Chapter 8. The link of privacy to homeland security is obvious—even the most passionate hardliners for security measures have a private life and would like to have privacy and freedom. Therefore, the community has always to be aware of security measure consequences and how to counterbalance as much as possible.

As human beings, everyone wishes certain privacy rights enabling them to live with freedom, dignity, and integrity. The basic goal of these wishes is to decide on one's own life, have the opportunity to freely move further in life, and be able invoke changes at a personal rate.

To protect these goals, many European member states, as well as the European Commission, defined privacy laws to protect citizens' privacy, integrity, and freedom. There are two basic concepts regulating privacy:

- The collector of data owns the data. The citizens have some rights to proper protection of these data against theft or publication (U.S. approach).
- The collected data still belong to a large extent to the citizen on whom this data have been collected. The citizen has the right of inspection, correction, and that the data are used only for the specific purpose they were collected for. (EU approach, with lots of variation in the various member states). Especially well protected must be data on personal beliefs, religion, political attitude, and medical data.

Security is based on the fact that control increases security. Typical control instruments are:

- Collecting data on communication (e.g., connection data, which must be stored for 6 months in Europe).
- Monitoring data communication and storing its content.
- Creating networked databases (NDB) correlating data of multiple-source databases, which were collected for various and different purposes. NDBs are the primary concern of EU data protection evangelists, because citizens' information are more and more transparent to national authorities.
- Checking individuals at the airport with a full body scan.
- Surveillance cameras.
- Positioning information of mobile phones.

A general fact is that observed individuals under surveillance feel less free. This has impact on their behavior, their creativity, and their mind. And that is the way privacy—in the long run—has impact on welfare and innovation as well: less privacy and freedom leads to less welfare and innovation.

Society, in general, and politicians in particular have responsibilities to elaborate a balanced mixture of security and privacy concerns, which is well tuned to the spirit of time, the thread and vulnerability of the situation, and to the public opinion.

As a fact about the consensus concerning the dilemma security-privacy, every security incident has impact and provokes reaction of responsible persons (for example, as politicians and leaders of corporations) such as the elaboration of new countermeasures to increase the security and preparedness level. Generally, within the window of opportunity after an incident, such measures are well perceived and digested by the public, and opposition is not really remarkable.

However, times change, and after a period the new accepted security measures have shifted the balance towards more control and surveillance, and less privacy. As an observation, we could make a thought experiment. Imagine the regulation and freedom of 1995 compared to today. Suddenly, we realize that the society has accepted many controls, which would not have been feasible outside the after incident window of opportunity. Moreover, it would not be feasible to build consensus on such controls in regular times.

This thought experiment discloses that the dilemma of security-privacy has a unilateral development at least in many European states. The forces to increase control with the hope of more security has a natural development. Meanwhile, the opposite process does not have a sufficiently strong lobby to maintain citizen rights, individual freedom, and integrity on a constant level. Concluding on this observation, the natural development towards a very strong surveillance society does not stop, but additional measures are introduced.

Suggestions for privacy and freedom-maintaining controls are:

- Use of privacy enhancing technologies, for example, a limited period of storing video camera streams, or authorizing a citizen for easy access of his data for inspection, correction, and—depending on the application—erasure. A number of European Commission–funded ICT-FP7 projects are addressing approaches for these important issues in more detail [e.g., Think-Trust (www.think-trust.eu/)], and a number are examining the impacts of these technologies. Some examples include: PrimeLife (www.primelife.eu/), PICOs (www.picos-project.eu/), TAS3 (www.tas3.eu/), and SWIFT (www.ist-swift.org/).
- Introduce periodic assessment of the security gain of each countermeasure with consideration of financial investment and operational cost, as well as the loss of privacy and freedom. The goal of assessment is to rejustify risk reduction of the countermeasure in a period, without particular pressure to react on actual incidents.

As indicated at the beginning of this section, society needs both security and privacy. Finding the right balance is not an occasional, but a permanent task, which should be carried out at least periodically.

3.6 Conclusions

The discussion in the preceding five sections has shown that homeland security and challenges in information systems have substantial dependencies and interrelations. Both homeland security and military defense has to include from now on information security and critical information infrastructure in their programs.

Catastrophes like in the nuclear power plant in Chernobyl, the 9/11 incident in New York, the recent financial turmoil, or volcanic eruption in Iceland may be minor when looking at the impact of a real breakdown of the cyberspace. Cyberspace and ICT is no longer a comfortable add-on into society—it is a very basic prerequisite of nearly any action in our everyday lives, either as a part of a general infrastructure or directly as an information infrastructure.

Nations urgently should assess risk in the critical information infrastructure, if not yet already done, and carefully follow the CII development re-evaluating risk situations. Natural human behavior and experience is in contradiction to ICT technologies, cyberspace, and its development:

- A generation of ICT lasts around three, but a maximum of five years.
- Every attack concern must be made on global scale. In physical space, noone expects intrusion of a criminal on the other side of the world, but in cyber crime, this is very likely.
- Physical systems are ideally made to degrade gracefully. In cyberspace attacks like the attack on Google's infrastructure (news.cnet.com/1606-2_3-50082324.html) or the potential attack on the U.S. energy grid, lead within minutes to unexpected situations. There is no such thing as graceful degradation in cyberspace.

The difference between police and military is well understood. However, the difference and interrelation of information security, critical infrastructure protection, and homeland security still needs a lot of communication until all decision makers understand and are able and willing to invoke the right countermeasures. Therefore, if not already done, nations should increase efforts now, before a computer Chernobyl happens.

References

[1] Brunner, E. and M. Suter, *International CIIP Handbook 2008/2009*, http://e-collection.ethbib.ethz.ch/eserv/eth:31095/eth-31095 01.pdf

[2] Dondossola, G., et al., "Evaluation of the Effects of Intentional Threats to Power Substation Control Systems," *Proceedings of the International Workshop on Complex Network and Critical Infrastructure Protection 2006, International Workshop on Complex Network and Critical Infrastructure Protection*, Frentari Center, via Frentani 4 Roma, March 28–29, 2006.

Selected Bibliography

Bouwmans, I., M. P. C. Weijnen, and A. Gheorghe, "Infrastructures at Risk," in A. V. Gheorghe, et al., (eds.), *Critical Infrastructures at Risk*, New York: Springer, 2006.

Rich, E., et al., "Emergent Vulnerability in Integrated Operations: A Proactive Simulation Study of Economic Risk," *International Journal of Critical Infrastructure Protection*, Vol. 2, No. 3, 2009, pp. 110–123.

Rinaldi, S. M., J. P. Peerenboom, and T. K. Kelly, "Criticalò Infrastructure Interdependencies," *IEEE Control Systems Magazine*, 2001.

Setola, R., and S. Geretshuber, "Critical Information Infrastructure Security," *Third International Workshop CRITIS* , 2008.

Sveen, F. O., et al., "A Group Model Building Approach for Identifying Simulation Scenarios in Critical Infrastructure," *43rd Annual IEEE Hawaii International Conference on System Sciences (HICSS-43)*, Kauai, HI, 2010.

References

Selected Bibliography

4

Analysis of Emerging Phenomena in Large Complex Systems

Alfonso Farina, Sofia Giompapa, and Antonio Graziano

4.1 Introduction

In these years, a holistic approach has prevailed in science (e.g., physics, biology, engineering, software, social sciences) and has also invested the domain of large and complex systems for homeland defense (HD) and homeland security (HS). Due to political, doctrinal, and economic issues, the concept of system in HD and HS has grown to encompass a larger ensemble of actors and systems which cooperate to provide an extended capability; *looking at the big picture* is at the same time an opportunity, since additional rationalization and increased effectiveness may be achieved and a necessity, due to existing interdependencies.

With reference to such systems, the term "complexity" is, sometimes, an abused word; other terms would often be more appropriate such as complicated or garbled. On the other side, even though the interest for the science of complexity is growing up in a very fast way, it is not easy to select a shared definition of "complexity," and it is difficult to define the boundaries of a complexity science. Nevertheless a consensus is being gained on the main characteristics that complex systems share: a complex system consists of multiple simple units, which interact nonlinearly and exhibit a behavior that is not merely given by adding up the behaviors of the individual components. A usual feature of these systems is that their components are not organized from the outside, but they are self-organizing.

This chapter attempts to clarify the emergent aspects of this new science of complexity in relation to large systems in the domains of HD and HS. The rest of the chapter is organized as follows. Section 4.2 shortly addresses the main features of complex systems and related challenges. Section 4.3 introduces the potential of new engineering approaches to the design of large complex systems. Section 4.4 describes three study cases related to large complex systems: the analysis of the congestion phenomenon of homogeneous and heterogeneous traffic in Internet-like networks, the domino effect in a large high voltage electric distribution grid, and the analysis of interdependencies in large critical infrastructures. Finally, concluding remarks are provided in Section 4.5.

4.2 The Challenge of Complex Systems

In the following we provide a very brief introduction to complex systems, focusing on those aspects that help to put in the right perspective large systems in the domains of HD and HS. For our purposes systems can be classified into three fundamental categories: simple, complicated, and complex [1]. Note that in relation to the scale of observation the same system may be classified in a different way. Simple systems have a small number of components that have well-defined roles and are governed by well understood rules. Complicated systems have a large number of components that have well-defined roles and are governed by well understood rules; in this kind of system robustness is achieved through redundancy. Complex systems, as mentioned in the previous section, typically have a large number of similar components, which may act according to rules that may change over time and that may not be well understood. The connectivity of the components may be quite plastic and roles may be fluid. In these systems robustness is achieved by enabling the parts to adapt to the changing environment and adopt different roles. The most distinguishing feature of a complex system is that it is capable of emergent behavior (i.e., a self-organizing collective behavior difficult to anticipate).

Complex systems can be further classified into complex nonadaptive systems and complex adaptive systems [2]. The former class includes those systems that are based on models comprising the immutable equations of physics (e.g., weather forecasting) or where rules of behavior are hard-coded into genes (e.g., micro-organisms and ecosystems). Their behavior may adapt to circumstances, but it does so in a predictable way, which is also encoded in their genes; they offer the promise of predictability. The latter class includes systems that are based on models and agents (e.g., people) that can adopt new rules when circumstances change (e.g., financial markets). In this case understanding the behavior of a complex system necessitates a simultaneous understanding of the environment of the system and conjectures are translated in phenomenological equations. These need to be complemented by a model of interactions between systems.

It should be considered that when systems grow in size, a cardinality challenge may also emerge [3]. Scale is the way that things change as the size of the problem is increased; things get bigger and bigger, and eventually the size of the problem crosses a line so that the increase in size fundamentally changes the problem. Polynomial complexity, which is usually considered a manageable issue, can become a significant problem when large scale data are taken into account; even something which grows very slowly, like sorting, becomes close to impossible at scale.

4.3 The New Approach of System Engineering

Engineering highly complex systems in uncertain environments requires a different approach with respect to traditional engineering of systems with well-defined inputs and outputs. Complex systems cannot be studied with the classical approach of the black box, by considering separately their input, output, and state, since the interactions between their components are fundamental in order to understand the overall behavior of the system. The classical methods of system theory, automatics, control theory, which are based on analytical studies of the stability and controllability of systems, are no longer the best way to evaluate the new complex systems, due to the strong interactions existing between their components [4, 5]. Nowadays complex large systems are no longer characterized by a vertical structure, such as in the past, but they are characterized by direct and indirect interconnections with other systems [6]. Multiple new factors complicate their analysis, such as technological and scientific factors, but also economic issues, social behaviors, regulations and normative aspects, and globalization of information technologies. All these factors were not decisive in the past, and therefore they were not considered in the classical studies of systems. The new trend of complex systems, which evolve according to a horizontal and branching structure, forces us to look for new analysis tools.

A useful tool, which can aid the comprehension of complexity elements in large systems and the interdependencies between structures, is modeling and simulation (M&S) [7]. A model is a simplified representation of a system during a certain time instant or at a certain point in space. It has the purpose to understand the essential aspects of a real system. A simulation is the manipulation of a model, which acts a compression of space and time and allows evaluating and forecasting the dynamic progress of a set of events, and observing interactions which are not otherwise observable, due to their separation in space and time [8].

An interesting example of a large complex system is represented by complex critical infrastructures, whose correct functioning is fundamental for the everyday operations of a whole country. The simulation of one or more interdependent infrastructures is related to different scales [9, 10]: spatial, geographical, and temporal scales that can be widely extended. The spatial scale of an infrastructure can be extremely wide, since it can deal with a single part, with an infrastructure or a metastructure, which include more infrastructures and the environment. The concept of geographical scale is related to that of spatial scale, since infrastructures can extend along physical spaces, which include more cities, regions, or nations. Finally, the temporal dynamic of the infrastructure can be a few milliseconds (e.g., power generation processes), several hours (e.g., fluid/gas transportation, and so forth), or several years (e.g., processes that are concerned with improvement of the infrastructure or the addition of more functionalities).

There are also operational factors that influence the behavior of large complex systems, which are difficult to model and then to insert into the simulation: (1) the presence of one or more human operators inside the control chain, the operator reaction time, and the aspects related to the training; (2) the security policies of the infrastructures; (3) the occurrence of unpredictable events, such as emergency situations or crisis, which involves elements of risk analysis and management [11, 12]. All these facts can alter the system behavior and the results of its simulation. One of the main challenges of complex systems is the ability to develop simulation structures that are specific enough and obey to the following requisites:

- To understand the mechanisms that produce a system and to understand how they can be included into a simulation;
- To be able to handle a large multiplicity of data, often having an heterogeneous nature and located on different sites;
- To comprehend the interactions between different models and to consider different analysis contexts.

The science of complexity is relatively recent and immature and a clear evidence of this is the fact that there is no accepted measure of complexity. Several possible measures have been proposed such as the computational complexity of a system, in terms of space and time; the logical complexity of a system; the complexity measured according to the typical quantities of information theory[1]; the complexity of one entity according to the regularity of its structure. Yet none of these is decisive and when no well-defined metrics are available, a consequential difficulty arises in the management of the system. This difficulty has been highlighted by the Nobel laureate Murray Gell-Mann [13, 14], who has been a promoter of the foundation of the Santa Fe Institute, in New Mexico, dedicated to the study of the science of complexity.

A further crucial element of the new study approach is the definition of validation measures for the models and the simulations, in addition to the definition of suitable metrics that allows evaluating the performance of a complex system. Such kind of metrics should take into account a large number of factors.

As it concerns a large complex system it is not easy to define a univocal set of measures for performance evaluation, since these measures are extremely dependent on the scenario, on the elements that compose the system and their interactions, and on the main functions and finalities of the considered system. It is possible to identify some criteria that associate these measures and can provide schematic indications for their definition. More specifically, some crucial points concerning the performance metrics of a large system are:

- The performance of the processes that take place in the system (e.g., information exchange, information/data fusion, communication processes, control processes, and so forth);
- The capability of the system to react to emergency situations (reaction typology, reaction time, reaction commensurability, and so forth);
- The analysis of the vulnerability, threat, and risk elements;
- The overall robustness of the system, its tolerance to damages (unintentional or intentional), and its recovery capability;
- The evaluation of the effectiveness of the security policies and the determination of strategies that can increase the global security of the system.

Measures developed according to the factors listed above are very useful in order to examine the results coming from M&S, and to compare them with potentially available real data.

The new trend of engineering forces and encourages us to design and exploit complexity to obtain additional performance and robustness for the systems. An appealing idea which is being pursued in several fields (e.g., network management, environmental monitoring, structural monitoring, and so forth) is to exploit emergent behaviors and to purposefully design emergent behaviors in systems. For instance, the study of autonomic network management is

1. As an example, the algorithm information content (AIC) is referred to the length of the more concise description that can be used for a certain entity [13].

gaining momentum, especially in the information technology (IT) domain. IT systems are expected to embrace a set of operations, such as [15]: autodiscovery (i.e., to recognize newly added elements and changes in existing ones); proactive performance trend monitoring, which takes into account changes or exceptions compared with normal variation in behavior; automated determination of root causes of problems to accelerate recovery; some degree of automated corrective actions. An interesting example of emerging behavior regards the dynamical autosynchronization of the elements in a network. To better understand this aspect lets consider the brilliant experiment of two or more metronomes placed on a base conceived by Pantaleone [16, 17] and shown in Figure 4.1.

In Figure 4.1(a) two metronomes on a table oscillate in a random fashion; the pendulums are not moving in rhythm with each other because they are uncoupled (i.e., they do not share a common pulse that directs them to move in a unified fashion). When the two metronomes are placed on a board that rests on two slightly rotating soda cans, as shown in Figure 4.1(b), they become coupled devices and, after an adjustment period, become synchronized. A common pulse is established and it creates a unified action.

Emergent behaviors could thus be exploited to increase performance of large complex systems, but probably technologies based on these behaviors are not mature enough. For this reason deeper studies and researches are required to better understand the complex interdependencies among the systems. The use of methodologies based on the development of models and on the implementation of simulation is certainly a useful tool for the growth of the science of complexity.

4.4 Study Cases

This section describes three study cases that have been conducted considering emergent phenomena in complex large systems. The first study case considers the phenomenon of traffic congestion in Internet-like networks. The second study case analyzes the domino effect in a large high-voltage electric distribu-

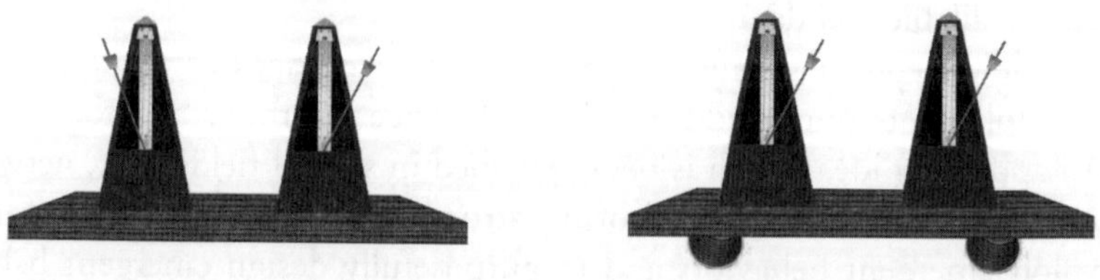

(a) (b)

Figure 4.1 Synchronization of metronomes: (a) asynchronous and (b) synchronous metronomes.

tion grid. Finally, the third study case deals with the analysis of the interdependencies in large critical infrastructure.

4.4.1 The Traffic Congestion Phenomenon in Internet-Like Networks

4.4.1.1 Network's Topological Properties

The congestion problem in network flows is of great interest in several engineering fields, such as, for example, pipeline flow, electric power transmission, telecommunications, highways, or railways traffic management. The network considered in this first study case is the Internet, which is characterized by a continuous growth in size. The choice of this specific network is due to the fact that it is crucial in everyday life and its possible congestion is a relevant issue from the social and the economic points of view. For this reason, this network can be an attractive target of an intentional attack aimed to deliberately overload its ordinary flow to cause the network congestion and its consequent collapse (as an example let consider a cyber war scenario).

The specific purpose of this study case is the analysis of the data packet traffic as a function of the topology and the load of the network, in the two cases of homogeneous and heterogeneous traffic. A homogeneous traffic network is characterized by only one kind of data packets moving on it, while a heterogeneous traffic network is characterized by more than one kind of data packets moving on it.

Models for transmission networks have been widely addressed in the literature [18–22], with special attention to the congestion problem. In this analysis the network is studied by means of the associated weighted graph $G(V, E)$, where $V = \{1, 2, \dots n\}$ is the set of n nodes/vertex, and E is the set of branches/edges. The graph is characterized by its topological features: the node degree distribution, the mean length of the shortest path between couples of (distinct) nodes, the graph connectivity and the betweenness centrality (BC), which for a given node v is the average percentage of the shortest paths connecting two nodes m and k ($m \neq v$ and $k \neq v$) and which cross the node v [23]. The BC is a measure of the relevance of the node for the traffic on the network when a shortest path is used as routing strategy in the network. The following quantities characterize the network graph: $e_{m,k} \in E$ (for $m, k \in V$ and $m \neq k$) is the branch that connects node m with node k; $c_{m,k}$ is the capacity associated to branch $e_{m,k}$ (the capacity $c_{m,k}$ is a positive integer and is the maximum number of data packets that can go through the branch $e_{m,k}$ in a time unit moving from the source node m to the destination node k); $p_{m,k} = 1/c_{m,k}$ is the weight associated to branch $e_{m,k}$.

In this study case two networks are considered as concerns the congestion phenomenon: Inet 3037 and a modified rectangular lattice. Inet 3037 is a scale-free[2] network with 3,037 nodes[3] that is supposed to reproduce the topological properties observed experimentally on the Internet and is generated by using the Inet-3.0 software package [24]. The modified rectangular lattice is a simple modification of a rectangular lattice, with the addition of an extra node such that it has the same number of nodes as Inet 3037. The two types of networks are represented in Figure 4.2.

Some topological properties of the Inet 3037 and the modified rectangular lattice are summarized in Table 4.1.

Table 4.1 shows that the nodes of the modified rectangular lattice have approximately the same relevance in terms of betweenness centrality, since the mean value of the betweenness centrality is of the same order of magnitude of its maximum value for the modified rectangular lattice. This is a consequence of the fact that the modified rectangular lattice does not have the small world property.[4] The case of Inet 3037 is substantially different—in fact the mean value of the betweenness centrality is three orders of magnitude smaller than its maximum value.

The analysis in this study case starts from the observation that complex networks (i.e., the networks having the small world property) and lattice networks behave very differently with respect to the congestion phenomenon. Having this in mind a transformation is defined [25], depending on a parameter

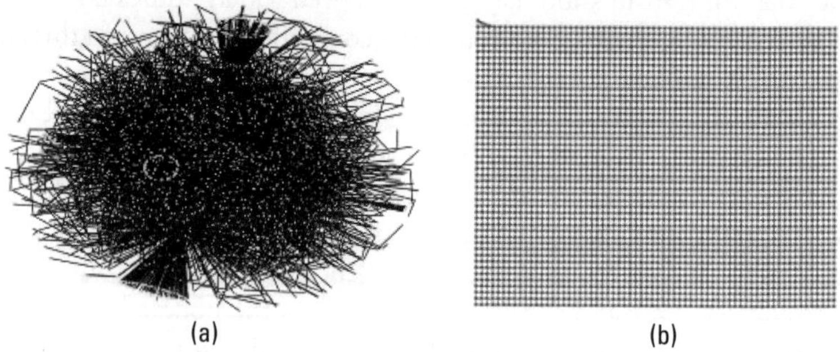

Figure 4.2 Networks considered in the study case: (a) Inet 3037 and (b) modified rectangular lattice.

2. Scale-free networks are characterized by a power law distribution of connectivities [i.e., the fraction $P(k)$ of nodes in the network having k connections to other nodes goes, for large values of k, as $P(k) = k^{-\gamma}$ where γ is a constant whose value is typically in the range $2 < \gamma < 3$].

3. 3,037 is the minimum number of nodes generated by the Inet-3.0 software and it also represents the number of autonomous systems on the Internet in November 1997.

4. The small-world property of a network implies that most nodes in the graph are not neighbors of one another, but most nodes can be reached from every other by a small number of hops or steps.

Table 4.1 Some Features of Inet 3037 and the Modified Rectangular Lattice

Feature	Inet 3037	Modified Rectangular Lattice
Mean length shortest path	4.6241	38.3424
Mean length shortest weighted path	5.1649	38.3424
Maximum value betweenness centrality	0.5785	0.0267
Mean betweenness centrality	$8.646 \cdot 10^{-4}$	0.012
Maximum value weighted betweenness centrality	0.4604	0.0267
Mean value weighted betweenness centrality	$1.00 \cdot 10^{-3}$	0.012

$\alpha \in [0,1]$ that maps a network having the small world property into a (modified) lattice network that has the same number of nodes. Let E^I denote the set of branches of Inet 3037 and $\mathbf{A}^I \in \mathbf{R}^{3037 \times 3037}$ denote the weighted adjacency matrix of Inet 3037 (i.e., the matrix whose entries are $A^I_{i,j} = c^I_{i,j}$, $i \neq j$ and $A^I_{i,i} = 0$, where $c^I_{i,j} = c_{i,j}$ if $e_{i,j} \in E^I$, $i,j = 1,2,\ldots,3037$. Let $\alpha \in [0,1]$ be a real parameter and let consider the following combination of the weighted adjacency matrices $\mathbf{A}^R$ and $\mathbf{A}^I$ relative to the modified rectangular lattice and to Inet 3037, respectively:

$$\mathbf{A}^\alpha = \left\lceil \left((1-\alpha)\mathbf{A}^I + \alpha \mathbf{A}^R \right) \right\rceil \alpha \in [0,1] \tag{4.1}$$

where $\lceil \cdot \rceil$ is the ceiling part of its argument (i.e., the smallest integer greater or equal to the argument). The matrix $\mathbf{A}^\alpha$ is the weighted adjacency matrix of a graph called α-network. Note that when $\alpha = 0$ the matrix $\mathbf{A}^\alpha$ is equal to the adjacency matrix of Inet 3037, and when $\alpha = 1$ the matrix $\mathbf{A}^\alpha$ is equal to the adjacency matrix of the modified rectangular lattice.

This mapping, which changes the capacity of the branches of the graphs, can be regarded as an "interpolation" between the scale-free network and the rectangular lattice. Using this transformation, the behavior of Inet 3037 ($\alpha = 0$) can be compared to the behavior of the modified rectangular lattice mentioned above ($\alpha = 1$) and the behaviors of the interpolating networks ($0 < \alpha < 1$) can be studied. In this study we intend to show how the change in the network topology can alleviate the congestion phenomenon.

4.4.1.2 Congestion Phenomenon in Homogeneous Traffic Case

The study of the congestion phenomenon on the network described by the weighted adjacency matrix (4.1) is first performed in the homogeneous traffic case. The congestion phenomenon is studied in stationary conditions and the following two quantities are considered: the mean travel time of data packets,

t_M, and the mean number of packets that have not reached their destination and are still traveling on the network, N_L.

In the homogeneous traffic case the following queue management rules are assumed: (1) the node queues are potentially unbounded and they can be empty; (2) the first in first out (FIFO) rule applies (i.e., the packet that arrives first is delivered first); (3) if two packets arrive simultaneously and there is enough capacity, they are delivered at the same time; (4) if two packets arrive simultaneously and there is not enough capacity, the first generated packet is delivered; and (5) if two packets arrive at the same time, there is not enough capacity and they have been generated at the same time, a random choice is made.

At time zero the networks studied are empty. This fact generates a transient behavior that connects the initial state to the large time behavior of the traffic on the network. The simulation of data packet traffic is carried out in the time interval $[0, T]$, where T is chosen equal to 400 time units to guarantee that in the free flow regime the traffic on the networks has approximately reached the stationary condition. The simulation conditions are listed here:

1. The number of network topologies that will be considered in the simulation is selected ($N_\alpha + 1$) and the network topologies are given by expression (4.1) where $\alpha = \alpha_j = (j - 1)/N_\alpha$, $j = 1, 2, \ldots, N_\alpha + 1$.
2. Packets are generated as samples from a Poisson distribution [25].
3. Routing is based on the shortest weighed path connecting the source node with the destination node.
4. A packet is removed from the simulation when it reaches its destination.
5. If there is not capacity available, the packet will wait for the next available time unit according to the queue management rules.

The parameters considered during the simulation are reported and defined in Table 4.2.

Figure 4.3 shows the quantities $N_L^\alpha(\lambda)/\lambda - t^\alpha$ and $t_M^\alpha(\lambda)$ as a function of λ for several values of the parameter α (i.e., for $\alpha = 0, 0.3, 0.5, 0.7, 0.9, 1$).

The quantity $N_L^\alpha(\lambda)/\lambda - t^\alpha$ is approximately constant up to a given value of λ and after that value of λ it starts to increase. The results shown in Figure 4.3 suggest that the networks with $\alpha \in (0.3, 0.7)$ are the best performing ones between those studied when the quantities N_L and t_M are used to judge the traffic quality. In fact in Figure 4.3 we can see that for $\lambda \in [100, 1{,}100]$ when $\alpha = 0.3, 0.5, 0.7$, the values taken by the quantities $N_L^\alpha(\lambda)/\lambda - t^\alpha$ and $t_M^\alpha(\lambda)$ are smaller than the corresponding values of the same quantities when α is equal to 0, 0.9, and 1. The choice of the parameter α in the interval (0.3, 0.7) defines

Table 4.2 Simulation Parameters in the Homogeneous Traffic Case

λ	Mean value of the number of packets generated in the entire network in a time unit.
n	Number of nodes in the network.
λ / n	Average number of packets generated by each node per time unit.
$N_L^{\alpha}(\lambda)$	Mean number of packets that are traveling in the network and have not reached their destination.
$t_M^{\alpha}(\lambda)$	Mean travel time of the data packet.
λ^*	Value of parameter λ that separates the free flow state from the congested state.

a set of networks that is an interesting compromise between Inet 3037 ($\alpha = 0$) and the modified rectangular lattice ($\alpha = 1$).

4.4.1.3 Congestion Phenomenon in Heterogeneous Traffic Case

Let now consider the congestion phenomenon in the heterogeneous traffic case, where heterogeneous refers to the assumption that three kinds of data packets are moving on the network. This choice is inspired by the fact that in real telecommunication networks three kinds of packets can be distinguished: data, video, and voice packets. The waiting time associated to each packet waiting at a busy node is based on the arrival order at the node and on the priority associated to each kind of data packet. For the packets of kind i this waiting time is multiplied by the priority factor p_i, for $i = 1, 2, 3$. The priority factors are chosen as follows: $p_1 = 3$ for the packets having highest priority (packets of kind $i = 1$), $p_2 = 2$ for the packets having intermediate priority (packets of kind $i = 2$), and $p_3 = 1$ for the packets having lowest priority (packets of kind $i = 3$). The queue management rule consists in the choice of serving first the packet for which the product of the waiting time and the priority factor is the largest. If this value is the same for two or more packets, the packet generated earlier is delivered first; if packets have also been generated at the same time, the choice is performed randomly. When two or more packets present at a node must continue their route along the same branch and there is sufficient capacity available on the branch, the packets go through the branch in the same time unit.

The simulation interval is the same of the homogeneous traffic case and the simulation procedure is similar to steps 1 to 5 of Section 4.4.1.2, except for the packet generation and the queue management. In particular, the parameter of the Poisson distribution for packet generation is $\lambda_i = \beta_i \lambda$, for $i = 1, 2, 3$, where λ is a positive parameter and β_i for $i = 1, 2, 3$ are nonnegative con-

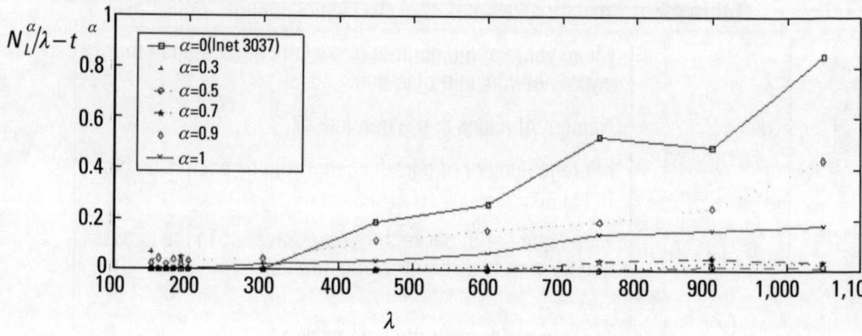

(a)

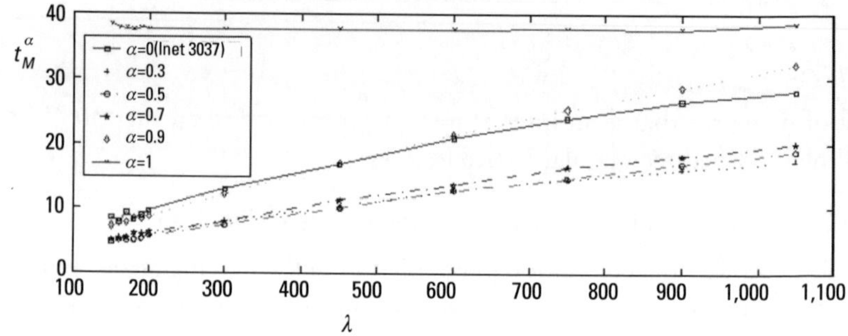

(b)

Figure 4.3 Performance of the homogeneous traffic case as a function of λ: (a) $N_L^\alpha(\lambda)/\lambda - t^\alpha$; and (b) $t_M^\alpha(\lambda)$.

stants such that $\sum_{i=1}^{3} \beta_i = 1$. The choice of the constants β_i determines different heterogeneous load conditions on the network.

The parameters considered in the simulation are those reported in Table 4.3.

The following values for the constants β_i, i = 1, 2, 3, have been selected: β_i = 1/6, β_2 = 2/6, β_1 = 3/6. This is the most common load situation, where the number of packets per unit time with highest priority (kind i =1) is (on average) smaller than the number of packets per unit time of intermediate priority (kind i = 2), and this second number is (on average) smaller than the number of packets per unit time with lowest priority (kind i = 3).

The parameters considered to evaluate the performance are: the mean travel time of data packets of kind i, $t_{M_i} = t_{M_i}^\alpha(\lambda)$, and the mean number of packets of kind i that have not reached their destination and are traveling on the network, $N_{L_i} = N_{L_i}^\alpha(\lambda)$. The results of the simulation are expressed in terms of the quantities $N_{L_i}^\alpha(\lambda)/\lambda - t^\alpha$ and $t_{M_i}^\alpha / t^\alpha$ as a function of λ, and they are reported in Figures 4.4–4.6, for α = 0 (Inet), α = 0.5, and α = 0.9999, respectively.

Table 4.3 Simulation Parameters in the Heterogeneous Traffic Case

λ	Mean value of the number of packets generated in the entire network in a time unit.
$\beta_i\lambda$	Mean load of the packets of kind i (i=1, 2, 3)
n	Number of nodes in the network.
$N_{L_i}^{\alpha}(\lambda)$	Mean number of kind i packets that are traveling in the network and have not reached the destination.
$t_{M_i}^{\alpha}(\lambda)$	Mean travel time for kind i data packet.
λ_i^*	Critical value of the load parameter λ for the α-network and the kind i data packet.

From Figure 4.4 and Figure 4.5, it can be observed that the kind 1 packets reach the criticality after the other kinds of packets when the Inet 3037 and the α = 0.5 network are considered. Figure 4.6 shows that in the α = 0.9999 network the priorities assigned to the different kinds of packets seem to have a small effect. In fact, when α = 0.9999 the values of $N_L^{\alpha}(\lambda)/\lambda - t^{\alpha}$ in the homogeneous traffic case and the values of $N_L^{\alpha}(\lambda)/\lambda - t^{\alpha}$ in the heterogeneous traffic case, for i = 1, 2, 3, are approximately the same. The effect of priorities in lattice-like networks (α = 0.9999, α = 1) is significant only for larger values of the parameter λ. In the free flow regime the traffic behavior on the α = 0.9999 network is very similar for the three different kinds and their travel times remain substantially the same as the travel time observed in the free flow regime for the homogeneous traffic case.

In conclusion, Figures 4.4 through 4.6 show that going from the α = 0 to the α = 0.9999 network the role played by the priority decreases. In fact, when α = 0 the curves corresponding to the three kinds of packets are well separated (i.e., the priorities are working), when α = 0.5 only the curve of the kind 3 packets is well separated from the remaining ones (i.e., priorities of the first two kinds of packets reduce their role) and when α = 0.9999, as already mentioned, there is no substantial difference in the three curves.

As a final remark, note how the network topology and the load condition (given by the choice of the parameters β_i) play an important role in the traffic congestion phenomenon in the heterogeneous traffic case.

4.4.1.4 Concluding Remarks on the First Study Case

The congestion phenomenon for the traffic of data packets in transmission networks has been studied as a function of the network topology and of the load conditions. The congestion of the network is a situation of interest in HS scenarios, since it can be the effect of an intentional attack aimed to overload

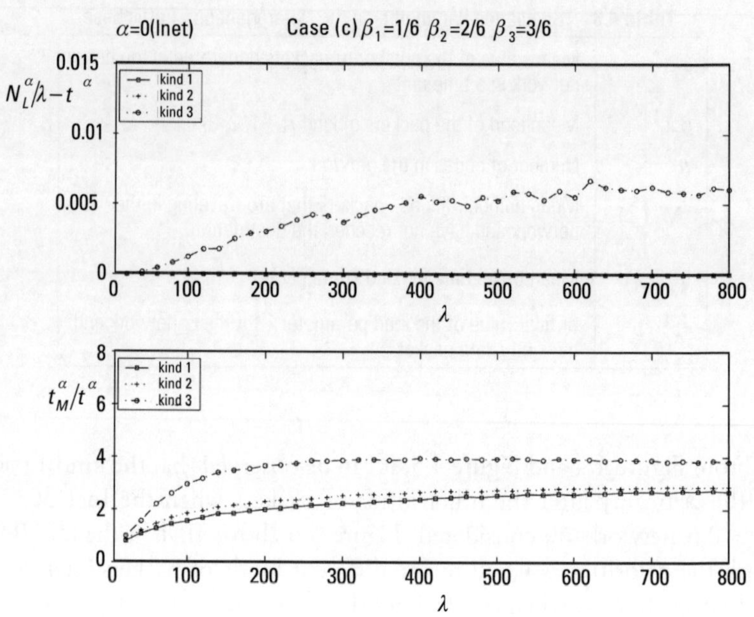

Figure 4.4 The quantities $N_{L_i} = N_{L_i}^{\alpha}(\lambda)$ and $t_{M_i}^{\alpha} / t^{\alpha}$ as a function of λ, when $\alpha = 0$.

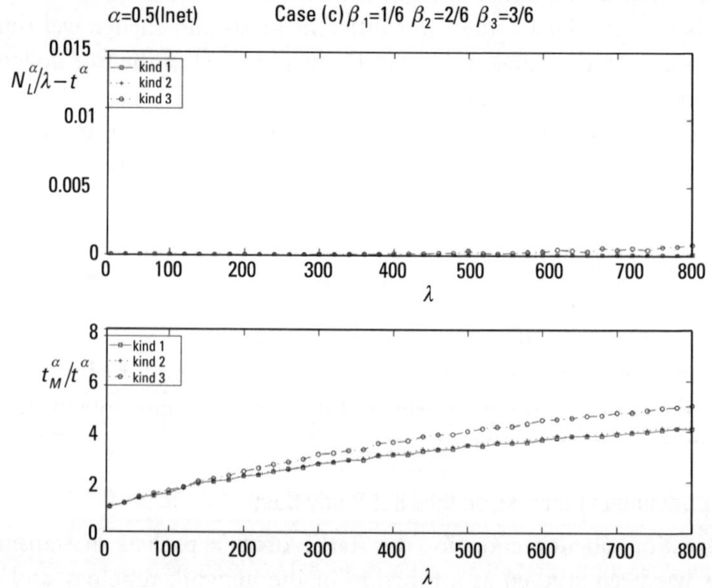

Figure 4.5 The quantities $N_{L_i}^{\alpha}(\lambda) / \lambda - t^{\alpha}$ and $t_{M_i}^{\alpha} / t^{\alpha}$ as a function of λ, when $\alpha = 0.5$.

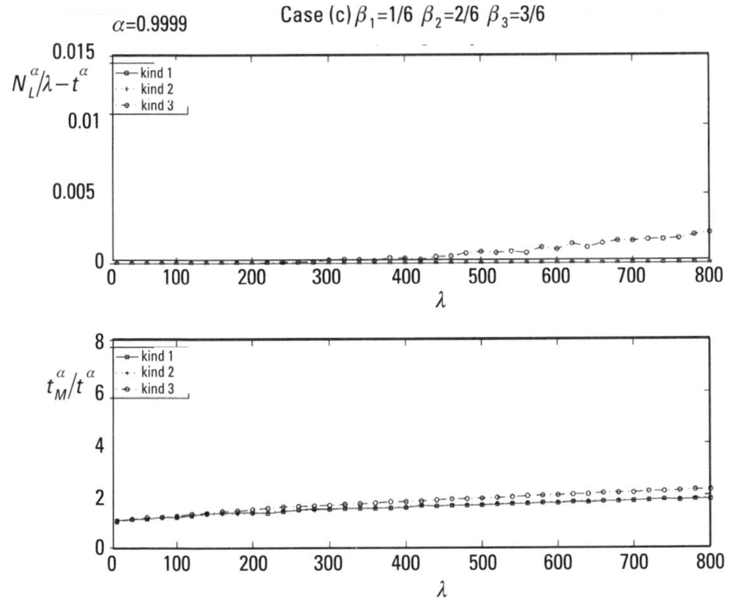

Figure 4.6 The quantities $N_{L_i}^{\alpha}(\lambda)/\lambda - t^{\alpha}$ and $t_{M_i}^{\alpha}/t^{\alpha}$ as a function of λ, when $\alpha = 0.9999$.

the network traffic and to cause its collapse, with serious social and economic consequences.

Two types of traffic have been considered: homogeneous traffic (i.e., when only one kind of packet is moving on the network), and heterogeneous traffic (i.e., when more than one kind of data packets is moving on the network). The congestion phenomenon has been studied under stationary conditions. Two parameters have been considered for the performance evaluation: the mean travel time of the packets in the network and the mean number of packets that have not reached their destination and are traveling in the network. Two kinds of networks have been presented: a scale-free network that emulates the Internet (Inet 3037) and a modified rectangular lattice network. The two networks have the same number of nodes. A transformation has been defined, which maps the network having the scale-free properties into the modified rectangular lattice. This transformation acts by changing the weight of the branch in the graph representing the network and it depends on the way the nodes of the network are numbered. The analysis conducted during the study case shows that network topology heavily affects the congestion phenomenon and suggests some actions on the traffic management rules and/or on the network topology to mitigate this phenomenon. In fact, there exist intermediate network topologies that are reasonable compromises between the network having the scale-free property

(Inet 3037) and the modified rectangular lattice: routing or topology adaptation is recommended to avoid traffic congestion. The topology of these intermediate networks is a simple interpolation between Inet 3037 and the modified lattice, obtained changing the capacities of the branches.

4.4.2 Domino Effect in a Large High-Voltage Electric Distribution Grid

4.4.2.1 Introduction to Blackout Dynamic

This section presents a study case concerning the failure analysis in a high voltage power transmission network [26] using the mathematical model of cascading blackouts introduced in [27].

Cascading failures are exhibited by large complex transmission networks. They consist in a sequence of failures occurring in a fast time scale as effect of the reduction of the operating margins in the transmission network due to congestions or line breakdowns. The interdependencies are the main reason of the vulnerability of power transmission networks and are translated in the risk of large blackouts. In fact, due to this interdependence, local overload of lines or local failures of components can generate cascades of breaks that, in extreme cases, can cause the fall of the entire network. This phenomenon is named cascading blackout, where blackout means the temporary absence of electric energy in some parts of an electric network. Thus the cascading blackout is a sequence of events characterized by line failures and/or load sheds.

The topic of cascading blackout is extremely interesting for HS scenarios. In fact, a sequence of failures causing a cascading blackout can occur as a consequence of both sabotage and an accidental malfunction. This can determine the total absence of service in an electric power grid, with a domino effect on other relevant networks and structures that depend on it, with deleterious consequences on everyday life.

The approach to the study of the blackout dynamic followed herein considers the network behavior in steady state conditions when random fluctuations are introduced in the load demand. The blackout dynamic is simulated as a sequence of equilibrium states and a statistical analysis of the blackout phenomenon is performed [27, 28].

4.4.2.2 The Optimal DC Power Flow Problem

A power transmission network can be modeled as an undirected graph, which is a set of nodes connected by branches. The nodes can be classified as generator nodes, load nodes, and junction nodes. The branches connecting the nodes represent the network transmission lines where the electric power flows.

In this study case, to examine the blackout dynamic, the network behavior in steady state conditions is considered, when random fluctuations are introduced in the load demand. In particular, the blackout dynamic is simulated as a sequence of equilibrium states and a statistical analysis of the blackout phenomenon is performed. This simulation approach, given a probabilistic model of the load demand, makes it possible to quantitatively study the reliability of the network. The approach has the advantage of being easy to deal with, but on the other hand it has the disadvantage of providing a crude approximation of the blackout dynamics. Assuming that the power transmission network operates in steady state conditions, the power flow along the network is described by a system of nonlinear algebraic equations and inequalities representing the network operation limits. The nonlinear equations are called power flow equations. In the statistical study performed herein, the power flow equations associated to random perturbations of the power demand are solved, but many different cases must be simulated to build a significant statistical sample of the network behavior. Therefore to simplify the numerical effort, a linearized version of the power flow equation is adopted, known in literature as DC power flow equations [29].

In normal conditions the upper limits on the power flowing along the network lines and on the power generated by the generator nodes are such that this set of constraints together with the DC power flow equations has a feasible solution. To simulate the cascading blackout dynamic, the network is overloaded by introducing random disturbance into the load nodes [27]. The introduction of this initial disturbance changes the request of power to the generators and, consequently, the power flowing along the lines. A sufficiently large initial disturbance can cause cascades of line breakdowns and/or of load sheds. In fact, when a line is overloaded and goes down, the power flowing along it is redistributed to the neighboring lines that unexpectedly must transport an extra flow and eventually become overloaded. This event can cause their fault. In the same way, when the request to a generator exceeds its power generation capacity, the other generators must supply, if possible, the extra power load demand. In both cases there is a redistribution of power flow in the network that must satisfy the DC power flow equations and constraints. When this redistribution is not possible the blackout dynamics start. In addition, when the system of DC power flow equations and constraints is feasible it does not have a unique feasible solution: for this reason, the power flow problem is reformulated by adding a cost function to the set of equations and constraints. This function measures the cost associated to the total power generation. The power flow distribution problem consists in minimizing, with respect to the power injected by the generators in the network, the cost function subject to the equality and inequality constraints representing the power balance law, the DC power flow equations and the upper bound limits of the power produced by the generators and by the line capacity (i.e., the maximum power flow that the line can tolerate). This constrained optimization problem is called optimal DC power flow problem. Since constrains have been linearized, a linear programming problem is finally obtained.

4.4.2.3 Cascading Blackout Model

The most frequent cause of blackouts in power networks is the increase of the load power demand. In this study case the cascading blackout dynamics are simulated as a consequence of variation of the load power demand P_D, assuming assigned the values of the remaining parameters. When the total load power demand P_D grows, the optimal DC flow problem and its solution go through some changes: the blackout occurs when the optimal DC power flow problem becomes unfeasible. The reason of this unfeasibility is the need to violate the upper bounds of the power produced by the generators and by the line capacity. In this case it is necessary to shed loads and/or to disconnect lines in the attempt of restoring feasibility of the model and to formulate a new problem. If the new problem is also unfeasible, new lines/loads are disconnected/shed until a feasible problem is found. These sequences of events generate the blackout cascade and mathematically correspond to formulate and to solve a sequence of constrained optimization problems.

A measure of the blackout size is the quantity P_S/P_D, where P_S is the undelivered power (i.e., the total power corresponding to the load shed). Note that $0 \leq P_S / P_D \leq 1$, and that $P_S / P_D = 1$ corresponds to a total blackout of the network.

There are analogies between the phase transition phenomenon of statistical mechanics and the behavior of a power transmission network in critical conditions (i.e., when, for instance, some lines are overloaded and/or generators cannot satisfy the power demand) [27, 30]. An example of phase transition in statistical mechanics is represented by the change of state of the water from liquid to gaseous state when the water temperature goes through 100°C at the pressure of 1 atmosphere. As the water temperature approaches 100°C, bubbles of water vapor (gas) appear in the liquid water; these bubbles contain a huge number of adjacent molecules that are adopting a collective behavior (i.e., these molecules are all in gaseous state). Similarly, in a power transmission network the total power demand can be interpreted as the analogous of the temperature in statistical mechanics and the increase of the total power demand can make the optimal DC power flow unfeasible. The successive line failures and/or load sheds, operated to make the optimal power flow problem feasible, can generate "bubbles" of blackout immersed in the power transmission network.

In order to study the phase transition phenomenon in the network, a statistical analysis of the blackout size can be performed [27, 28]. In particular, to overload the network a random disturbance of known statistical properties is added to the components of the load power vector: in this way the total power demand randomly changes and the resulting blackout size P_S/P_D becomes a random variable. In this case, the quantity P_D represents the mean value of the total power demanded by the network when the loads are randomly perturbed. The statistical analysis consists in generating a sample of the random variable P_S/P_D, for a given value of the mean P_D, and in finding an approximation for the probability density function (PDF) of the blackout size P_S/P_D. The study of this PDF as a function of P_D gives the required information about the phase transition phenomenon in the network. In realistic networks for reasonable values of P_D, the probability of having large blackouts is small, while blackouts of a small dimension are much more likely than large blackouts. Thus the PDF of the blackout size P_S/P_D decreases when P_S/P_D increases. The PDF of the blackout size can be fitted with a negative exponential or with an inverse power law [26]. The PDF of the blackout size associated to a power transmission network below the phase transition (i.e., P_D sufficiently small) is best fitted by a negative exponential when P_S/P_C increases (P_C being the set power): in fact when P_D is small, the possible load shed due to the fluctuations does not influence the other loads, since the network has a large margin of operation. As the power demand P_D increases, due to the increased power demanded to the generators, even a small fluctuation of the power demand may cause a blackout cascade. Therefore the decay of the PDF of P_S/P_D above the phase transition (i.e., P_D sufficiently large) is best fitted by an inverse power law function of P_S/P_D, when P_S/P_D increases. The phase transition phenomenon consists in this change in the behavior of the PDF of P_S/P_D. The critical value of the total power demand, P_D^*, is the value for which, when $P_D/P_C = P_D^*/P_C$, the PDF of P_S/P_D changes from being best fitted by a negative exponential function to being best fitted by an inverse power law. In the power transmission networks of practical interest this transition point may not exist and in any case it is difficult to establish its existence numerically.

4.4.2.4 Results of the Study Case

In this study case a statistical analysis of the random variable P_S/P_D is performed as a function of the mean total power demand P_D in the case of a high voltage power transmission network. An approximation of the PDF of P_S/P_D is obtained via simulations, each comprising 20,000 runs, for increasing values of P_D. A regression analysis is then performed on the approximated PDF and two functions are exploited to fit the data: the negative exponential and the inverse power law. It is postulated [26] that the PDF of P_S/P_D decays following an exponential law when P_D is less than a certain value $P_D^{(1)}$, while the function that best fits the data is the inverse power function when P_D is higher than the $P_D^{(2)}$, where it is assumed $P_D^{(1)} < P_D^{(2)}$. Consider now the parameters σ_e^* and σ_p^*, which minimize the errors deriving by approximating the PDF with an exponential and an inverse power function respectively. It is therefore expected that $P_D^{(1)}$ and $P_D^{(2)}$ exist (with $P_D^{(1)} < P_D^{(2)}$) such that when $P_D < P_D^{(1)}$ it is $\sigma_e^* \leq \sigma_p^*$

and when $P_D > P_D^{(2)}$ it is $\sigma_e^* \geq \sigma_p^*$. The evaluated quantities σ_e^* and σ_p^* versus P_D/P_C are plotted in Figure 4.7.

Figure 4.7 shows that for P_D/P_C values up to 0.74, the PDF is best fitted by an exponential law (i.e., $\sigma_e^* < \sigma_p^*$); for $P_D = 0.77P_C$ and $P_D = 0.81P_C$, the PDF is best fitted by an inverse power law (i.e., $\sigma_e^* > \sigma_p^*$). This analysis connects the phase transition phenomenon in a power transmission network to the appearance of fat tails (inverse power law behavior) in the PDF of the blackout size measure. The presence of these fat tails means that the probability of having large blackouts is significant above a certain critical value of P_D. In the studied high voltage power transmission network, in the configuration specified by the considered snapshot, the critical value P_D^* lies in the interval P_D^1, P_D^2, that according to Figure 4.7 can be chosen equal to $[0.74P_C, 0.77P_C]$.

Finally, Figure 4.8 shows the mean values of the line occupation (i.e., the ratio between the power flow through the line and the line capacity when $P_D/P_C = 0.74$).

From Figure 4.8 it is possible to see that the line occupation is not homogeneous: the occupation of some lines (lighter lines) is significantly greater than the occupation of the remaining lines (darker lines) of the network. This unbalance is the cause of the possibility of large blackouts when the mean value of the total power demand is smaller than the total network capacity; the lines whose occupation is close to one (light line in the figure) are the most vulnerable in the considered network. The reason of this vulnerability is related to the

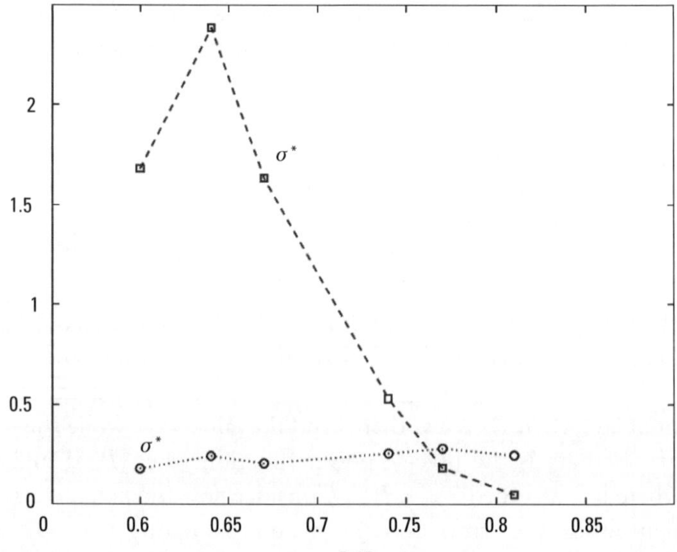

Figure 4.7 σ_e^* and σ_p^* versus P_D/P_C.

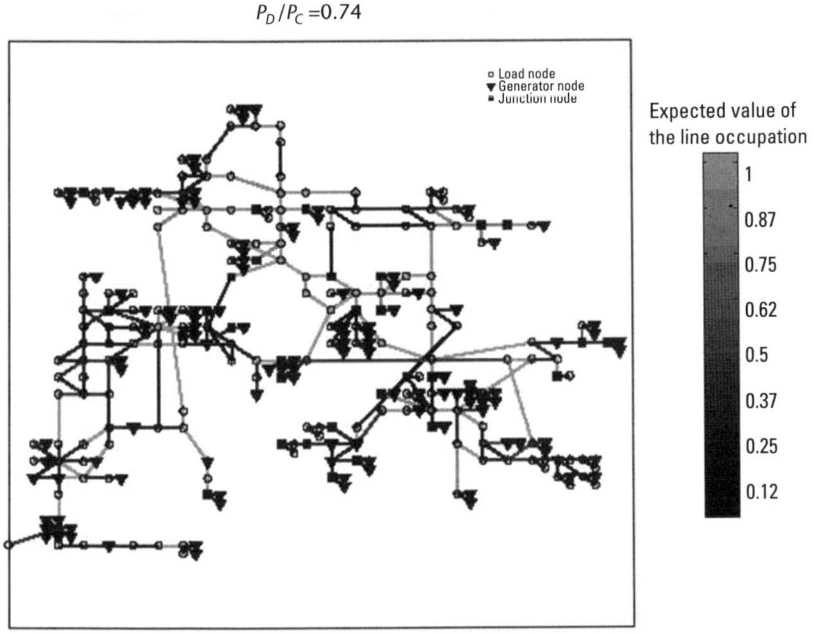

Figure 4.8 Mean value of the line occupation for $P_D/P_C = 0.74$.

network topology [31] and, in particular, to the connectivity properties of the graph representing the network.

4.4.2.5 Concluding Remarks on the Second Study Case

A failure analysis has been conducted on a high voltage power transmission network using the mathematical model of cascading blackouts. The blackout phenomena in a high voltage power grid and the PDF of the blackout magnitude have been studied by means of a statistical simulation. The analysis points out the existence of critical values of the mean total power demand such that the PDF of the blackout size changes from being best fitted by a negative exponential to being best fitted by an inverse power law. Since the inverse power law has a fat tail, this implies that beyond the critical values the blackout probability becomes significantly higher than that expected for an exponential model.

4.4.3 Interdependence Analysis in Large Critical Infrastructures

4.4.3.1 The Sociotechnological Scenario

Examples of critical infrastructures are the systems for production, transmission, and distribution of electrical power, telecommunication systems, water systems, transportation systems, banking and financial services, and so forth. These infrastructures are indicated as critical since damage to one of them can have a serious effect on the health, safety, and wellness of the citizens and on the correct operation of the government [32–34].

The protection of the national critical infrastructures is a topic that has recently gained interest and today it is one of the main concerns for many governments. In the past the infrastructures used to be autonomous and they held few contact points. Nowadays infrastructures are interconnected and mutually dependent in a complex way: a phenomenon that affects one infrastructure can have a direct or indirect impact on other infrastructures, spreading on a wide geographical area and affecting several sectors of the citizen life. This is schematically represented in Figure 4.9 [34].

The interdependencies between critical infrastructures increase their level of complexity, since they introduce multiple connections, feedback or feedforward paths, and thus they produce connected and branching topologies.

4.4.3.2 Interdependencies: The New Vulnerability

Interdependencies between infrastructures may be induced by a plurality of causes, many poorly known or completely hidden. The presence of them can augment the threat and possible targets (both direct and indirect), can amplify the consequence of a failure and allow simultaneous destruction of more than one service, and also can introduce unforeseeable behavior. Due to these reasons new and dangerous vulnerabilities can affect the entire infrastructural system [35, 36]: a failure, accidental or intentional (e.g., terrorist attack), can spread in an unpredictable way and affect a large number of users. These considerations

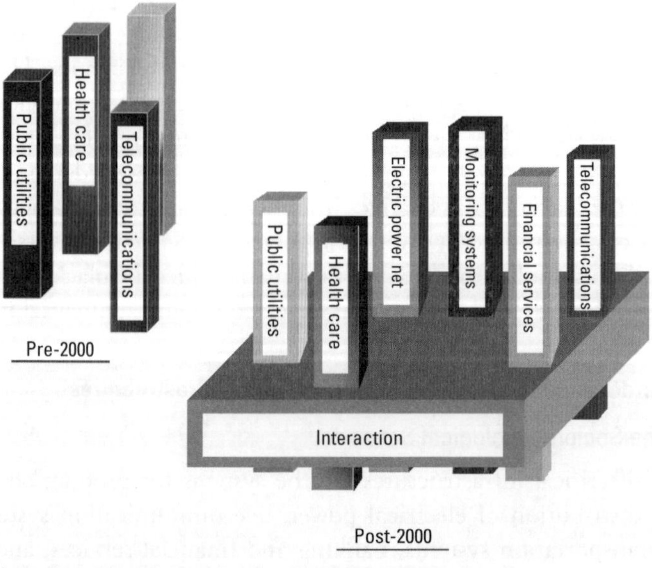

Figure 4.9 Interdependencies between present infrastructures.

have induced governments and international organizations to focus their attention to the improvement of safety and security, robustness, and resilience of such infrastructures [37]. The current scenario requires consideration of coordinated and transversal approaches, in addition to the classical strategy for security specific for each application area. These new approaches are indicated as critical infrastructure protection (CIP) or critical information infrastructure protection (CIIP), when the focus is on the information communication and technology (ICT) components of the systems [37, 38].

4.4.3.3 The Models of the Infrastructure

The considered study case analyzes a portion of a coastal region, as shown in Figure 4.10.

The functional and topological architectures of the infrastructures located in the selected region have been considered and the impact, in terms of degradation in the essential services, of an intentional attack or a technical failure that damages the entire technological infrastructures have been evaluated.

In the following, we report a synthetic description of the models for each infrastructure considered in the study case.

- *Communications:* Both the wired and the wireless networks have been considered. The analysis has concerned the segment between the concentration points located in the territory and the access gateway to the telephone backbone. This choice is due to the fact that the redundancy of the optical packet network makes the network robust enough against a local event, while the modeling of the last mile is not interesting due to the limited effect and to the impossibility to acquire the relative data.
- *Electrical grid:* The high voltage portion of the grid (220 kV and 380 kV) and the generation level have been considered. The distribution segment has been neglected due to its high tangle and to the difficulties in collecting data relative to the distribution network.

Figure 4.10 Coastal region considered in the study case.

- *Railway network:* The railway sections existing in the selected region have been considered and the functional relations between the railway infrastructure and the other modeled infrastructures have been taken into account.
- *Harbors:* The mission of the harbor as a connection hub between the maritime and the land transportation has been considered.
- *Airports:* The airport function as a hub between the air and the land transportations has been considered and its interactions with the other infrastructures have been modeled.
- *Highway network:* The highway sections that are placed on the analyzed area are considered in terms of services that they provide to the other infrastructures (people/good transportation) and in terms of services that they require from the other infrastructures.
- *Urban area:* The urban areas have been considered as resource providers (urban services, turnover personnel), and also as users of resources provided by the other infrastructures. One major urban area and five smaller urban areas present in the selected region have been considered.

4.4.3.4 The Simulator Model

A software simulator, called Critical Infrastructure Simulation by Interdependent Agents (CISIA), has been exploited to evaluate in a qualitative and quantitative way the consequences of a potential attack or an unintentional failure. The CISIA simulator [39] has been developed by University Campus Bio-Medico and University Roma Tre, both in Rome, Italy. The software language adopted is C++ and the simulator consists of about one hundred thousands lines of code. CISIA adopts a modular and abstract representation for the different components of the infrastructures to allow a consistent description even though incomplete and generic data are available. More specifically, the entities inside CISIA are macroscopic high level models, obtained by means of the knowledge derived by interviews to professional experts in the relevant domain (as shown in block A of Figure 4.11). These macroscopic models allow CISIA to be very general and to adapt to a wide range of situations. In the study case considered here, approximately 100 interviews have been made to develop the entity models of the simulator. CISIA primarily represents the dependences and interdependencies between the different components of the scenario, since these are the main causes of complex and unpredictable behavior. Several mechanisms have been modeled in CISIA, which encode different proximity concepts, and highlight the typologies of reciprocal influence between components.

The simulator is composed of two main structures: the entity pool, which manages the interior dynamic of each entity, and the transmission subsystem (TSS), which manages the different interactions between entities [40]. The scheme of the CISIA simulator is shown in block B of Figure 4.11. CISIA has been designed as a set of entities that evolve according to a temporal scanning defined by calculus cycles, which interact by means of a plurality of mechanisms (encoded through the nets shown in block B of Figure 4.11). Each entity is implemented by the class SimEntity (i.e., a simulation entity) that coordinates the execution of these calculus cycles. More generally, one scenario is characterized in terms of entities and adjacency matrices and it is implemented by the object SimEntity, which contains two subsystems:

- *Entity pool (EP)*: A collection of interdependent entities;
- *Transmission subsystem (TSS):* A system for the management of messages.

The simulation provides a large quantity of data; for this reason an appropriate user interface has been developed, which is able to provide the synthetic simulation data together with cartographic information obtained by means of Geographic Information System (GIS) techniques [41]. The architecture of the entire simulation process is schematically represented in Figure 4.11. The GIS provides the cartographic support and allows visualizing the entities of interest and their interconnection on the map of the area of interest together with other relevant information; a three-dimensional representation and an interactive navigation over the area of interest (flyover) allow the users to capture the evolution of the situation, providing a significant support to the final operators.

4.4.3.5 Study Case in Details

The study case described in this section is merely notional and no real data have been considered during the analysis. Additional approximations and alterations have been introduced, in order to make the simulation results unfit for inappropriate uses. This does not alter the goodness of the simulation results since the purpose of the study is the validation of the analysis methodology and of the simulator.

The study case considers a coastal region, which extends inland for approximately 5 km. The area is represented in Figure 4.10. The geographical area taken into account includes:

- A major urban area;
- Other minor urban areas;
- Two airports (a main airport and a secondary airport);
- Two different railways paths;

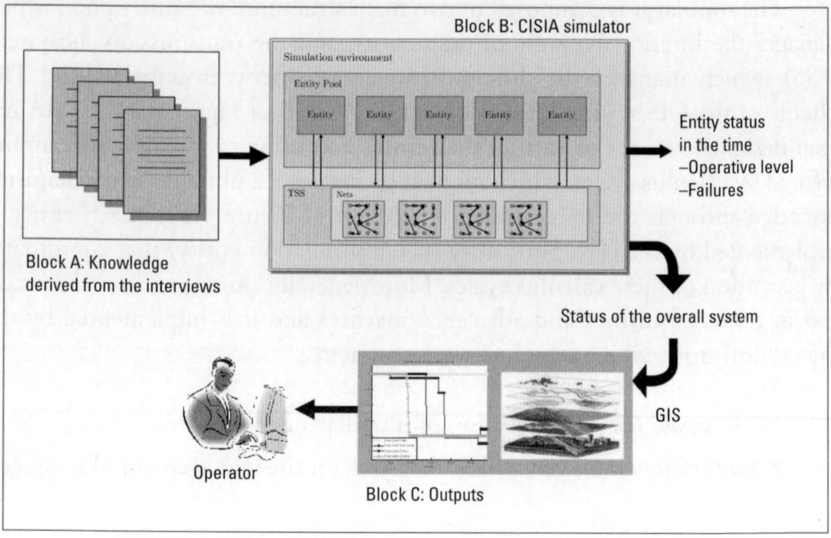

Figure 4.11 Scheme of the simulation process.

- High-voltage transmission lines and four electric plants;
- A complex telecommunication system;
- A harbor area interconnected with the electrical plants and with the transportation infrastructures;
- A second harbor area;
- Several important highways.

The graph that models these entities and their interconnections is composed of 233 nodes and 844 branches. The area has been considered for a time interval of 48 hours, thus the functional relations whose execution time is higher than the entire duration of the study case have been neglected.

During the study case different impact analyses have been performed, considering both isolated and sequences of events. The simulated scenario assumes the occurrence of an event (sabotage or accidental failure) that causes a damage to an access gateway (GTWA[5]) of the telecommunication network in an urban center near the coast. The GTWA represents the connection element as it concerns the entire data transmission between the users and the national telephone backbone. Therefore its inactivity implies the obstruction of

5. A GTWA is the central element of the telecommunications infrastructure. It permits the supply of wireless and data services.

all data connections, and the inactivity (both for data and phone calls) for all the providers that use the national telephone backbone.

It has been assumed that the failure event occurs at the time $T = 120$ minutes. One hour after the event, there are not appreciable effects on the normal activities in the area under test. At the time $T = 280$ minutes there is a gradual degradation in the operations of the train stations near the affected point. It is worth to remark that this degradation is not related to the railways signaling, since these kinds of signals use owner networks and thus they are not affected by a failure in the GTWA. Instead the problems that occur in the train station are related to the incorrect activity of the ticket machines (due to the missing connection with the central servers for the management of the reservations, the payments by credit cards, and so forth) and to the message/information exchange between passengers. The percentage of degradation in the operative level (i.e., the ability of an entity to provide the 100% of its nominal output capacity) is described in Figure 4.12 for the train stations. The sampling time in Figure 4.12 is equal to 10 minutes.

From Figure 4.12 it can be noticed that the operative level, referred to on the vertical axis between 100 (top level) and 0, decreases with the time and it has a propensity to become equal to zero. The degradation also regards stations that are physically far from the geographic point where the failure occurred (e.g., station A in Figure 4.12). Due to the incorrect activity of the train stations, the mobility in the area moves towards the car traffic. As a consequence, thirteen hours after the failure event ($T = 900$ minutes) there is a decreasing of the operative level of the highways infrastructure, as illustrated in Figure 4.13 in terms of percentage of traffic level, that is related to the speed of the car flow in the highway. The traffic level is referred to on the vertical axis between 100 (top level of traffic level) and 0. Again the time is sampled with a sampling interval equal to 10 minutes. Note that in Figure 4.12 and in Figure 4.13 only some of the curves obtained during the simulation have been reported, to simplify the visibility of the figures, and the markers on the line are not indicative of the sampling interval.

The problems that occurred to the train and highway traffic influence the access ways to the two airports of the considered area; as a consequence, a difficulty in the airport operations occurs. Note that the degradation to the operative level of the airports is due to the difficulties to the auxiliary services rather than to the air traffic itself, since this type of traffic owns dedicated channels and circuits. In fact, the reduced mobility in the entire area affects the passenger's inflow/outflow, the turnover, and the supply of different auxiliary services (e.g., catering).

Finally, the current operative level of the two airports is shown in Figure 4.14. The results of the simulation have shown that the reduced mobility also impacts the operative level of the two harbors in the area.

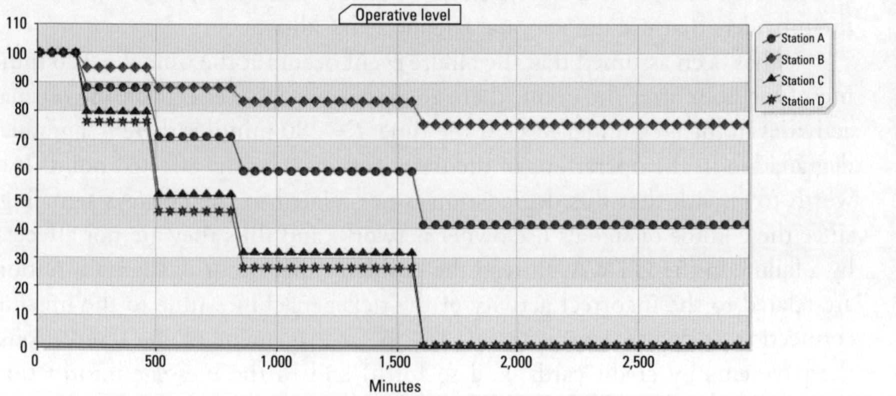

Figure 4.12 Operative level of the train stations versus time [min].

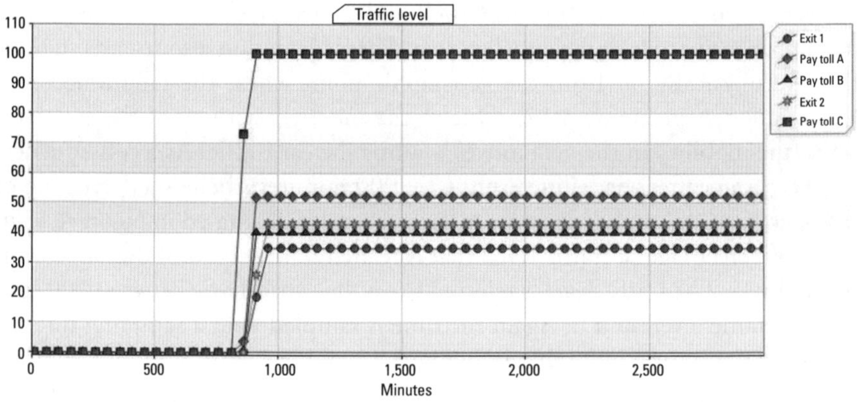

Figure 4.13 Traffic level in highway infrastructure versus time [min].

To conclude, at the end of the simulation (i.e., 48 hours after the beginning) the failure to the GTWA causes significant deterioration levels not only to the telecommunication network, but also to the railway, highway, airport, and harbor infrastructures. This is basically due to the degradation of the mobility capability in the entire considered area, even far from the failure point, and to the long persistence of the failure itself.

4.4.3.6 Concluding Remarks on the Third Study Case

As stated above, the considered scenario does not claim to represent the reality of the geographical area considered. In fact, the data in the study case have been

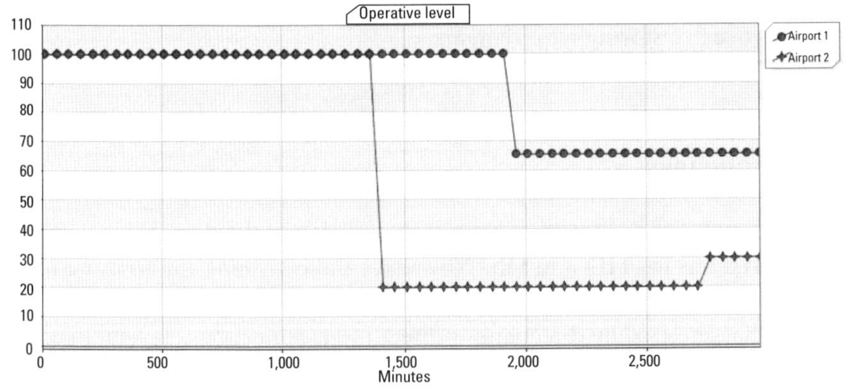

Figure 4.14 Airport operative level versus time [min].

intentionally altered in order to make them inapplicable for unlawful purposes. However, this alteration has not modified the complexity of the scenario and the architectural consistence of the real situation, and thus has not influenced the aims of the conducted analysis.

Note that the results of this study case are influenced by the assumption of absence of recovery phenomena; in a real context, a failure scenario, such as those considered in this study case, would have activated recovery/restoring operations.

The results are also influenced by the following fact: all the quantities considered in CISIA are fuzzy numbers with a triangular representation. For the sake of simplicity, the graphics reported in Figures 4.12–4.14 refer to the main value of these fuzzy numbers (i.e., the value with a higher reliability). To better understand the results, the other components of the fuzzy number should be considered, in addition to the main value.

As a final remark, the developed study case highlights the importance of modeling and simulation [7] in the analysis of interdependencies in complex phenomena scenarios: the CISIA tool allows the users to consider in a unique and complex model all the different interdependent infrastructures that exist in a certain geographical area. More specifically, tools like CISIA can be useful to conduct two kinds of investigations:

- To detect the critical points of the infrastructure in order to understand where to focus the protection and redundancy measures;

- To understand the critical factors of the infrastructures to appropriately assess the backup elements, the relationships with the providers, and the recovery/restoring requirements.

4.5 Conclusions

A holistic approach is spreading in science and also in the domain of large and complex systems for HD and HS. Nowadays these systems are no more just a collection of mostly independent vertical structures, such as in the past, but they are characterized by direct and indirect interconnections. Several new factors complicate their analysis: huge and evolving networks, technological and scientific factors, but also politics, the trend of finance, social behaviors, regulation and normative aspects, and the globalization of information technologies. Due to these reasons a new approach to complexity is required to better understand the behavior of these systems and consequently new analysis tools must be considered.

In this chapter we have tried to address the analysis of large systems by taking inspiration from the science of complexity and complex systems. Specifically, we have addressed three study cases. The first study case considers the phenomenon of traffic congestion in Internet-like networks and shows that the topology of a large network and routing algorithms severely influence its behavior. This study case gives evidence for the effectiveness of a topological approach to the analysis of large systems. The second study case concerns the analysis of the domino effect in a large high voltage electric grid. In particular, it is shown how cascading blackouts, that are a sequence of events characterized by line failures and/or load sheds, are exhibited by large complex transmission networks. In this case the interdependencies are the main reason of the vulnerability of power transmission networks and are translated in the risk of large blackouts. This second study case shows that a large network exhibits phase transition and domino effects, which are typical of complex systems (e.g., phase transition arises in liquid-vapor transition in statistical mechanics). Finally, the third study case deals with the analysis of the interdependencies in large critical infrastructures. More specifically, it describes how a failure in a component of the telecommunication network propagates its damaging effects on the other infrastructures present in the same geographical area: the railway network, the highway network, the airports, and the harbors in the considered area. Also this study case illustrates how a modern network of infrastructures is characterized by several interdependencies, known and unknown, and this introduces new global and often unpredicted vulnerabilities, which are typical of complex systems.

We have shown the importance of taking into account techniques and results from the science of complexity in the study of large systems and the potential of a new engineering approach that helps mastering and exploiting complexity. The theory of complexity indicates great promise in understanding and managing large complex systems with reference to domains of interest to us, such as HD and HS. This concept is well expressed by Stephen Hawking [42] in the following citation: "I think this century will be the century of complexity."

Acknowledgments

The authors would like to acknowledge all the colleagues of the Finmeccanica Working Group on the analysis of large systems. We especially thank Professor Francesco Zirilli, University Sapienza, Rome, Italy, for the cooperation on Sections 4.4.1 and 4.4.2; Professor Roberto Setola, University Campus Bio-Medico (Rome, Italy) and Professor Stefano Panzieri, University Roma Tre, Rome, Italy, for the cooperation on Section 4.4.3; and Professor Giorgio Franceschetti, University Federico II, Naples, Italy, for his accurate review of the chapter.

References

[1] Amaral, L. A. N., and J. M. Ottino, "Complex Networks. Augmenting the Framework for the Study of Complex Systems," *The European Physical Journal B*, Vol. 38, 2004, pp. 147–162.

[2] Woods, J. D., "Primitive Equation Modeling of Plankton Ecosystems," in *Ocean Modeling: Conceptual Basis and Applications*, N. Pinardi, and J. D. Woods, (eds.), Berlin, Germany: Springer-Verlag, 2003.

[3] scienceblogs.com/goodmath/2008/11/scale_how_large_quantities_of.php.

[4] DiMario, M. J., "System of Systems Interoperability Types and Characteristics in Joint Command and Control," *Proc. of the 2006 IEEE International Conference on System of Systems Engineering*, Los Angeles, CA, April 24–26, 2006, pp. 236–241.

[5] Bellomo, S., et al., "Attributes of Effective Configuration Management for Systems of Systems," *Proc. of the 2008 IEEE International Systems Conference*, Montreal, Canada, April 7–10, 2008.

[6] Noble, D., *The Music of Life: Biology Beyond Genes*, Oxford, U.K.: Oxford University Press, 2006.

[7] Farina, A., et al., "Large Systems Modeling and Simulation," in *Homeland Security Technology Challenges: From Sensing and Encrypting to Mining and Modeling*, G. Franceschetti, and M. Grossi, (eds.) Norwood, MA: Artech House, 2008, pp. 241–278.

[8] Bellinger, G., "Modeling & Simulation, an Introduction," http://www.systems-thinking.org/modsim/modsim.htm, November 2008.

[9] Rinaldi, S. T., "Modeling and Simulating Critical Infrastructures and Their Interdependencies," *Proc. of the 37th Hawaii International Conference on System Sciences*, Hilton Waikoloa Village, HI, January 5–8, 2004, pp. 1–8.

[10] Rinaldi, S. M., J. P. Peerenboom, and T. K. Kelly, "Identifying, Understanding, and Analyzing Critical Infrastructure Interdependencies," *IEEE Control Systems Magazine*, Vol. 21, No. 6, December 2001, pp. 11–25.

[11] Flegar, T., et al., "Protecting Cyber Critical Infrastructure (CCI): Integrating Information Security Risk Analysis and Environmental Vulnerability Analysis," *Proc. of the 2004 IEEE International Engineering Management Conference*, Singapore, October 18–21, 2004, pp. 888–892.

[12] Adar, E., et al., "Risk Management for Critical Infrastructure Protection (CIP) Challenges, Best Practices & Tools," *Proc. of the 2005 First IEEE Workshop on Critical Infrastructure Protection (IWCIP '05)*, Darmstadt, Germany, November 3–4, 2005.

[13] Gell-Mann, M., "What Is Complexity?" *Complexity*, Vol. 1, No. 1, 1995, pp. 16–19.

[14] Gell-Mann, M., *The Quark and The Jaguar: Adventure in the Simple and in the Complex*, New York: W.H. Freeman and Company, 1994.

[15] Hunter, P., "Network Heal Thyself," *Engineering & Technology*, Vol. 4, No. 14, August 8, 2011, pp. 58–60.

[16] Pantaleone, J., "Synchronization of Metronomes," *American Journal of Physics*, Vol. 70, No. 10, October 2002, pp. 992–1000.

[17] Ulrichs, H., A. Mann, and U. Parlitz, "Synchronization and Chaotic Dynamics of Coupled Mechanical Metronomes," *Chaos*, Vol. 19, 2009, pp. 1–6.

[18] Kocarev, L., and G. Vattay, *Complex Dynamics in Communication Networks*, Berlin, Germany: Springer-Verlag, 2005.

[19] Aiello, W., F. Chung, and L. Lu, "Random Evolution in Massive Graphs," in *Handbook of Massive Data Sets*, J. Abello, P. M. Pardalos, and M. G. C. Resende, (eds.), Norwell, MA: Kluwer Academic Publisher, 2002, pp. 97–122.

[20] Fabrikant, A. L., E. Koutsoupias, and C. Papadimitriou, "Heuristically Optimized Trade-Offs: A New Paradigm for Power Law in the Internet," *Proc. of the 29th International Colloquium on Automata, Languages, and Programming 2002*, P. Widmayer et al., (eds.), Lecture Notes in Computer Science 2380, Berlin, Germany: Springer-Verlag, 2002, pp. 110–122.

[21] Barabasi, A. L., and R. Albert, "Emergence of Scaling in Random Network," *Science*, Vol. 286, 1999, pp. 509–512.

[22] Barabasi, A. L., *Linked: How Everything Is Connected to Everything Else and What It Means for Business, Science, and Everyday Life*, New York: Plume Books, 2003.

[23] Newman, M. E. J., "Mathematics of Networks," in *The New Palgrave Encyclopedia of Economics*, 2nd ed., L. E. Blume and S. N. Durlauf, (eds.), Basingstoke, U.K.: Palgrave Macmillan, 2008, pp. 1–12.

[24] Jamin, S., and J. Winick, "Inet-3.0: Internet Topology Generator," *Technical Report CSE-TR-456-02*, Electrical Engineering & Computer Science EECS Department, University of Michigan, 2002, http://topology.eecs.umich.edu/inet/inet-3.0.pdf.

[25] Farina, A., et al., "Homogeneous and Heterogeneous Traffic of Data Packets on Complex Networks: The Traffic Congestion Phenomenon," *Networks*, 2011.

[26] Farina, A., et al., "Probabilistic Analysis of Failures in Power Transmission Networks and 'Phase Transition': A Study Case of a High Voltage Power Transmission Network," *Journal of Optimization Theory and Applications*, Vol. 139, 2008, pp. 171–199.

[27] Carreras, B. A., et al., "Critical Points and Transitions in an Electric Power Transmission Model for Cascading Failure Blackout," *Chaos*, Vol. 12, 2002, pp. 985–994.

[28] Carreras, B. A., I. Dobson, and D. E. Newman, "A Loading Dependent Model for Probabilistic Cascading Failure," *Probability in the Engineering and Informational Sciences*, Vol. 19, 2005, pp. 15–32.

[29] Andersson, G., Modelling and Analysis of Electric Power Systems, Lectures 35-526, Zurich: ITET ETH, http://www.ech.ee.ethz.ch/downloads/academics/courses/.

[30] De Marco, C. L., "A Phase Transition Model for Cascading Network Failure," *IEEE Control Systems Magazine*, Vol. 21, 2001, pp. 40–51.

[31] Bologna, S., L. Issacharoff, and V. Rosato, "Influence of the Topology on the Power Flux of the Italian High-Voltage Electrical Network," http://www.giacs.org/files/wp3_files/V%20Rosato%20Europhys%20Lett%20preprint.pdf.

[32] U.S. Government, *The National Strategy for The Physical Protection of Critical Infrastructures and Key Assets*, The White House, Washington, D.C., February 2003.

[33] E.U. Commission, Green Paper on a European Programme for Critical Infrastructure Protection COM(2005)576, Brussels, 2005.

[34] Protezione delle Infrastrutture Critiche Informatizzate – La realtà italiana, Presidenza del Consiglio dei Ministri – Dipartimento per l'Innovazione e le Tecnologie, March 2004 (in Italian).

[35] Zimmerman, R., "Decision Making and the Vulnerability of Interdependent Critical Infrastructure," *Proc. of the 2004 IEEE Conference on Systems, Man and Cybernetics*, The Hague, the Netherlands, October 10–13, 2004, pp. 4059–4063.

[36] Aijo, J., et al., "How to Define the Condition of an Infrastructure for Different Asset Management Purposes?" *Proc. of 2006 IEEE Conference on Systems, Man and Cybernetics*, Taipei, Taiwan, October 8–11, 2006, pp. 2629–2633.

[37] Dunn, M., and I. Wigert, *International CIIP Handbook*, Zurich: ETH, The Swiss Federal Institute of Technology, 2006.

[38] Gruppo di Lavoro Finmeccanica sull'analisi dei sistemi integrati per homeland security, SELEX Sistemi Integrati, *Sintesi dei risultati della cooperazione fra le Aziende del gruppo Finmeccanica*, Internal Report, Rome, Italy, September 2007 (in Italian).

[39] http://www.dia.uniroma3.it/autom/LabRob/Projects/CISIA/, April 2010.

[40] Panzieri, S., et al., "An Agent Based Simulator for Critical Interdependent Infrastructures," *Second International Conference on Critical Infrastructures*, Grenoble, France, October 25–27, 2004.

[41] Farina, A., A. Graziano, and E. Spinogatti, "L'integrazione delle competenze come chiave di successo per affrontare il problema della sicurezza delle Infrastrutture Critiche," *Safety & Security*, Maggio, 2010 (in Italian).

[42] Hawking, S., *Complexity Digest*, Vol. 10, March 5, 2001.

5

Model-Based Design of Trustworthy Health Information Systems

Janos Mathe, Jan Werner, and Janos Sztipanovits

5.1 Introduction

Healthcare is a knowledge-intensive activity. It is not surprising that twenty-first-century healthcare is rapidly moving towards a health care delivery system that is based on advanced information technology in virtually all aspects of the enterprise: laboratory tests, patient management, medical records, patient communication and interaction, and financial and administrative functions. Motivated by the promise of reducing preventable errors in patient care and minimizing administrative burdens, healthcare organizations (HCOs) have been migrating from traditional, paper-based records to health information systems (HISs), a collection of computer-based applications that enables sophisticated services for patients and health care providers. Examples for HISs include systems for electronic maintenance and exchange of health information, support of patient-centered decision making and a wide range of tools enabling effective communication between patients and providers. Various empirical evidence indicates that HISs can decrease healthcare costs [1–4], strengthen staff productivity [5, 6], and promote patient safety [7, 8]. As a consequence, HCOs are adopting HISs to enable a wide array of functions, including data sharing, decision support, employee training and student education, research, and access to reference materials. Many HCOs are leveraging HISs to build Web portals, which can be tailored to provide a specific experience based on the role of the intended user [9]. For example, physician portals have been designed to support daily clinical workflows, such that they enable access to guidelines, educational materials, treatment and cost information, and referral directories [10]. Similarly, patient portals have been designed to provide patients with access to their electronic medical records, billing, and appointment scheduling [11, 12].

Disparate HCOs, as well as departments within an HCO, can differ greatly in terms of the software and technologies that comprise their HISs. However, their design and implementation raises nontrivial challenges particularly in areas related to privacy:

1. In HISs, information flows are complex, dynamic, and evolve continuously driven by the rapid progress in medical knowledge. Following this evolution without risking the trustworthiness (dependability, security, and privacy) of the complex information systems requires advanced software and systems technologies in a rather conservative, highly safety conscious environment.
2. In traditional business applications, the human-computer interaction is often based on an accept or deny schema, where a person serves as an approval checkpoint that determines if a procedure or transaction can continue. On the contrary, in the health care domain, many information processing workflows require complex human tasks that do not fit this schema, such as a physician's interpretation of laboratory test results.
3. A third challenge is that in typical business applications, policies that impose restrictions on information flows across business units are not represented as independently manageable abstractions. Rather, they are frequently "hardcoded" in the workflow logic. To manage intricate and dynamic privacy policies common in the health care delivery environment, a more elaborate solution is needed.

There are extensive, multifaceted research efforts addressing these challenges both in the United States and in Europe [13–21]. Figure 5.1 captures the notional structure of the main research areas in the Team for Research in Ubiquitous Secure Technologies (TRUST) NSF S&T Center [22] targeting the development of trustworthy HIS.

1. Patient-centered clinical information management process research develops operational HISs, monitors operational experience (using logs and records obtained within hospitals), and extracts high-level abstractions and policies that are the starting point for improving and adapting key aspects of information services.

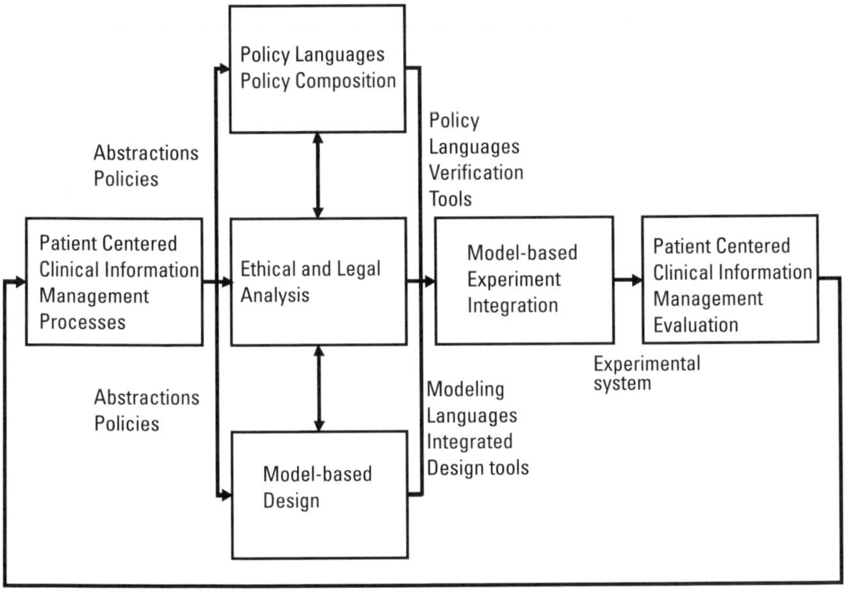

Figure 5.1 Notional structure of the main research areas in the TRUST Center, targeting the development of trustworthy HIS.

2. Policy languages and policy composition seeks to develop a general framework for modeling information flows and privacy policies in clinical environments that routinely use sensitive information. The research is built on insights gained by operational experience and seeks modeling methods that are able to address the problem of designing and verifying information flows to achieve both privacy and utility requirements. The results of this research activity are policy languages and verification tools.
3. Ethical and legal analysis focuses on the formulation and analysis of specific policies that reflect both societal norms and requirements of the effective enrichment and utilization of health information. The policies are expressed and composed using languages and methods that are precise and suitable for effective enforcement.
4. Model-based design targets the development of a scalable methodology for the modeling and model-based integration of HIS. The concept of "model" in this context simply refers to an engineering approach that evolves designs using formally defined, precise abstractions. This technique enables controlling design complexity by supporting the manipulation and integration of models for multiple design aspects (information flows, human organizations, software deployment, ac-

cess control mechanisms, and others) and by mapping high-level, domain-specific abstractions into standard-based modeling formalisms using model refinement and synthesis techniques.

5. Model-based experiment integration focuses on the rapid creation and integration of experimental HIS implementations. The primary challenge is heterogeneity of information systems in care delivery environments that makes interoperability a major challenge.
6. Patient-centered clinical information management evaluation focuses on the improvement of quality of health care by means of introducing advanced information technology health care delivery processes. A unique aspect of HISs is that IT penetration in care delivery needs to be measured in terms of this quality improvement and not by counting the number of IT devices or assessing the size of IT investments. It is a common requirement that HISs are evaluated in clinical trials to measure their effectiveness in the delivery environment.

In the following we discuss approaches in the research areas. In the discussion we do not seek for completeness. Our goal is to demonstrate the research directions and show their interrelationship (see Figure 5.1).

5.2 Patient-Centered Clinical Information Management Processes

HIS technologies offer providers and patients with the opportunity to operationalize, view, and contribute to medical records. However, the integration of electronic medical records in the treatment process (by patient management systems) and in the communication between patients and providers (by patient portals) create complex policy and technology management issues. These issues are complicated by conflicting federal policies on privacy and security for electronic medical records. The Privacy Rule of the Health Insurance Portability and Accountability Act (HIPAA) endows U.S. citizens with, among other privileges, a right of access to their medical records, the right to request amendments or corrections to those records, and the right to request an accounting of disclosures of their personal health information (HPPR) [23, 24]. At the same time, the HIPAA Security Rule (HPSR) requires healthcare organizations adopt physical, technical, and administrative security protections to record and monitor access to person-identifiable health information. The mandate of the Privacy Rule to provide patients access to their records favors the creation of more fully featured patient portals as a cost-effective method for disclosure and access audits; however the optimal design of such resources to protect confidentiality and minimize "information toxicity" is an unsolved research challenge. Below we present two examples for patient-centered clinical information management processes: patient portals and patient management systems.

5.2.1 Patient Portals

Since the mid-1990s, the National Institutes of Health have supported the development of prototype telemedicine systems for secure patient access to electronic medical records [25, 26]. The PCASSO system developed by Masys and his colleagues [27] provided patients and physicians with online access to medical records, the ability to audit access to records, messaging between patients and physicians, and authorization for healthcare functions. The users of these systems reported positive effects on care and no security breaches were reported in the portal systems. Yet, the pilot studies enrolled relatively small numbers of patients and were limited in their functionality. These systems may not be predictive of the societal effects of much broader availability of online medical records data. As patient portal technologies increase in functionality and become further integrated with the healthcare system, the information associated with a patient's medical record will become more complex and the potential for violations will become more diverse. Vanderbilt University Medical Center (VUMC) has been one of the pioneers in using patient portals effectively. The MyHealth@Vanderbilt portal is currently serving over 150,000 patients.

The preservation of privacy and security is paramount to ensure the growth and adoption of patient portals. The definition of privacy and security violations is contingent upon the policies set in place for the handling of patient health information. These policies need to specify who is permitted to access whose medical records, the parts of said records, and under what circumstances. For such scenarios, role-based access control (RBAC) is frequently used for data protection and is a requirement of the HIPAA Security Rule. As such, many off-the-shelf EMR systems are now equipped with RBAC. However, polices for RBAC are difficult to define and enforce in healthcare for a number of reasons, including:

- *Patient control:* Though patients wish to exercise control over access to their medical records, seldom do they understand the implications of their decisions.
- *Subjective views of privacy:* The privacy of personal health information is based on beliefs and is context dependent. These beliefs are not shared by all individuals with access to the same medical records [28]. People with dissimilar perceptions of risk and benefit manage access to shared health information in conflicting ways.
- *Complex interpersonal and organization roles:* Role-based access control schemes assume simple pair-wise relationships (e.g., provider patient or provider-office patient), but the reality of modern healthcare is that it is team based on the provider side and social network based on the patient side. This is further complicated by the fact that there are many groups

of individuals, with shifting roles in healthcare, each of which have legitimate access to an individual's health information.

Improving access control policies is the subject of active research. Without clear policies in place, many healthcare organizations use a "break the glass" model: they entrust care providers with broad access privileges, but stress that harsh punishments will result for misuse or abuse of the system.

Patient portals offer access not only to electronic medical records but a suite of services, such as messaging, appointments, and billing. Implementing these services require complex workflows that route information among patients and various agents of providers (doctors, nurses, clerks). While these workflows can be effectively implemented in the *service-oriented architecture* (SOA) framework [29, 30] and can be formally modeled using standard workflow modeling languages such as BPEL [21], ensuring the consistency between workflows and privacy rules remains an open issue and active research area [31].

5.2.2 Patient Management Systems

An emerging category of HISs is patient management systems that merge patient monitoring, decision support, and reporting/logging functions in an integrated platform. An example for patient management systems is Sepsis Treatment Enhanced Through Electronic Protocolization (STEEP) [32] that monitors real-time patient data streams and, using specific laboratory and vital signs criteria, identifies patients with possible sepsis and assist the providers (physicians, nurses) with presenting relevant information and advice for treatment choices. STEEP is a guideline-driven system. The sepsis patient manager is the software interface that shows the current state of sepsis management variables and provides management suggestions based on the information in the medical record and evidence-based guidelines. The clinical protocol that drives the recommendations was formulated after reviewing similar efforts at other institutions, reviewing the available medical literature, and with the consultation of various domain experts.

Evidence-based guidelines are represented as formal models. STEEP follows the treatment process by continuously interpreting the guidelines as new information from the patient and the provider personnel arrive.

Privacy and security considerations are central in the design of patient management systems. Safety concerns require not only validated guideline models but also demand their verification for critical properties, such as deadlock freeness. Similarly to patient portals, patient management systems are also subject to privacy regulations. The formal representation of guidelines that capture all information flows among actors of the treatment processes offer an opportunity for testing the consistency with privacy rules.

5.3 Policy Languages and Policy Composition

Enterprises that collect and use personal information face the challenge of effectively conducting their business without creating undue privacy risks. Nowhere is this more apparent than in healthcare and medicine, where it is unacceptable and unlawful to publicly disclose information about a patient's condition that could be vital to safe and effective diagnosis in a medical emergency. In order to improve privacy and security aspects of HISs such as patient portals in a way that supports planned future expansion of the portal, there is a need for a general framework for modeling workflows and privacy policies of enterprises that routinely use sensitive information. The reason for considering workflow and privacy together is that many privacy requirements (including notification requirements could be satisfied trivially if personal health information is not used in HIS activities. The challenge, therefore, is to design HIS services and activities and such patient portal services and patient management system activities so that useful tasks can be completed in ways that manage privacy risks effectively.

An essential part of the challenge is to develop policy languages and enforcement mechanisms for distributed and centralized and access control and privacy. From 1999–2004, Mitchell at Stanford University developed the RT role-based trust management system [33–38]. Trust management is an approach to distributed access control and authorization based on the following concepts:

- An access control policy establishes which parties have rights to access or delegate access to applications and information.
- A deduction engine determines, using distributed policy, whether to grant or deny access to a given resource.

In many approaches, signed certificates are used to authenticate policy statements and access requests, although other mechanisms are also possible. Policies determine when an entity or members of some group may perform an action on some object, such as a document, database, or purchase order. Deduction determines the consequences of a policy and produces proofs that can be enforced locally by access monitors. The RT system consists of a policy language framework and a set of deduction algorithms. RT is logically based, using Datalog as the foundation for the core language constructs and their implementation. Some of the more interesting features of RT are support by an extension of Datalog [39] with types and constraints, which leverages a significant body of work on constraint logic programming and constraint databases.

In recent years, the Stanford group has studied privacy policy languages and has developed an improved and more expressive approach to privacy based on Helen Nissenbaum's theory of contextual integrity. Contextual integrity is a conceptual framework for understanding privacy expectations and their implications developed in the literature on law, public policy, and political philosophy. The Stanford group has formalized some aspects of contextual integrity in a logical framework for expressing and reasoning about norms of transmission of personal information. In comparison with access control and privacy policy frameworks such as RBAC, EPAL, and P3P, they have found that many important privacy norms center on what the personal information is about, how it is transmitted, and what are the past and future actions by both the subject and the users of the information. Their proposed formal privacy model based on contextual integrity is expressive enough to naturally capture many notions of privacy found in legislation, including those found in HIPAA, COPPA, and GLBA. An advantage of the temporal logic approach to contextual integrity is that a number of important problems regarding compliance with privacy norms, future requirements associated with specific actions, and relations between policies and legal standards reduce to standard decision procedures for temporal logic.

The relationship between workflow and guideline models and privacy is that workflows and guidelines represent the functional goals of the organization, (such as the need to respond to requests for medical advice, or the need for sending specific medical data to provides for evidence-based decisions), and privacy imposes constraints on the disclosure of information to individuals inside and outside the organization. Current research at Vanderbilt University considers the representation of workflows and guidelines abstractly by logical formulas [40] and composes them with privacy constraints in a semantically solid manner.

A further important problem the Vanderbilt group investigates stems from the fact that medical and personal information is handled both by humans and by electronic information systems. As part of the design of a patient management or workflow process, human agents are given responsibilities for how to handle personal information. For example, a nurse might be responsible for answering health questions. While information systems can impose constraints on access to tagged information, the tags identifying the topic or scope of information in files or messages are under the control of individuals who input or modify information, and therefore can be inexact. Since this limits the accuracy of automated privacy policy enforcement, the development of auditing methods that can identify a set of individuals who might have violated workflow prescriptions and caused privacy breaches to occur [41].

5.4 Ethical and Legal Analysis

Progress in HISs must be evaluated beyond the level of services provided by specific systems, such as patent portals, and the novelty of the underlying technologies. HISs must be assessed with respect to the legal, ethical, and social policies that are pervasive in health care systems. Healthcare consumers will not take advantage of new health-related technologies if their personal, sensitive information is handled below an expected level of privacy and confidentiality. The inappropriate disclosure of sensitive health information can cause embarrassment, denial of insurance, or limit job prospects. Without trust in the system, health consumers will refuse to use advanced communications tools, or may withhold information that is pertinent for proper and timely diagnosis or treatment.

Other health consumers may refuse to use the system entirely, fearing medical identity theft, in which criminals gain access to personally identifiable information in order to impersonate that patient to gain care. In that instance, a victim of medical identity theft must deal with fraudulent charges, as well as lasting repercussions because the individual's medical files are altered to include the medical incidents of the impersonator. Protecting privacy and confidentiality must be a paramount consideration if new technologies and networked systems such as patient portals are to be successfully developed, deployed, and adopted by the public at large.

The legal and ethical rules and guidelines that govern the more traditional, nonelectronic communications between doctor and patient are only the starting point for the legal and ethical rules and guidelines for Web-based electronic communications. To that end, designers of health information technology and systems need to understand the crucial role that privacy and confidentiality play in health care systems. By identifying the safeguards that must be built into health information systems and networks from the outset, technologists can incorporate policy considerations into every aspect of the data collection, processing, and distribution process [42–44]. In doing so, these systems can help preserve patient privacy and confidentiality while ensuring that personal information remains accurate, credible, and consistent with the needs of health care providers and patients. As noted earlier, major policy statutes include the HIPAA Privacy Rule (45 CFR Parts 160 and 164), which provides minimum federal standards of privacy protection for health information, while allowing more stringent state laws. Information related to substance abuse and chemical dependency treatment is protected by section 543 of the Public Health Service Act, and its implementing regulation, 42 CFR, Part 2. This regulation, which supersedes both HIPAA and all more permissive state laws, requires that any disclosure of information related to substance abuse and chemical dependency treatment be accompanied by the individual's signed authorization.

Medical systems and organizations may also be subject to other federal laws with respect to privacy, including the Privacy Act of 1974, Family Educational Rights and Privacy Act, Freedom of Information Act, Employee Retirement Income Security Act of 1974, Gramm-Leach-Bliley Act, federally funded health programs regulations, Food, Drug and Cosmetic Act, Clinical Laboratory Improvement Amendment, federal disability and nondiscrimination laws, and U.S. Safe Harbor Privacy Principles.

Finally, 35 states now have Security Breach Notification Acts. In general, these laws attempt to address the growing problem of identity theft, which can include medical identity theft, by encouraging companies to be more vigilant about the security of personal information, and to be more open about security breaches that may have compromised consumer data. While the laws vary from state to state in terms of triggering events and overall applicability, most notification laws require businesses that control electronic data to notify state residents if their personal information is believed to have been acquired by an unauthorized person. These statutes often contain timing and content requirements.

Policies play two different roles in the design of HISs. First, policy-driven access control, privacy, and security requirements necessitate architectural solutions for system implementation. Here, the role of policy requires a formal and machine intelligible representation of policies that provide the basis for system verification and drive system operation, as we described in Section 5.3. Second, policies go beyond compliance with federal and state statutes to encompass the organizational, cultural and social norms of the system provider and the system users. Therefore, beyond the consistency between the technical and architectural choices mentioned in Section 5.3, the organizational, cultural, social, and legal interpretation of policies necessary to enable widespread adoption of HISs is an important research topic.

5.5 Model-Based Design

While research areas described in Sections 5.2, 5.3, and 5.4 provide abstractions, theoretical, ethical, legal foundations for patient-centered HISs, model-based design focuses on the formal representation, composition, and manipulation of the integrated models of information processes and security/safety policies and provides the tools for automated system generation directly from the models. It addresses layered, multiple view system modeling, model transformation, model analysis and validation, synthesis of implementations, execution, and design evolution.

An example for model-based design frameworks that was used successfully in creating HISs is Model-Integrated Computing (MIC) [45–48]. MIC is an approach and a tool suite for model-based software and systems engineering that have been developed over the last two decades at Vanderbilt. This infrastructure offers new opportunities in creating clinical decision support and process management systems. MIC focuses on formal representation, composition, and manipulation of integrated models of information processes and security/safety policies, and provides tools for automated system generation directly from the models. The open-source MIC tool suite [47] enables layered, multiple-view system modeling, model transformation, model analysis and validation, execution, and design evolution. The application of MIC principles and tools casts the creation of HISs in the following steps:

1. *Design of modeling language for application-specific workflows and guidelines.* In MIC, modeling languages are formally defined by metamodels [48]. The MIC metaprogrammable tools, for modeling, model management, and model transformation are automatically customized by the metamodels. The metaprogrammable infrastructure enables the selection of abstractions that are highly customized to a clinical environment or treatment process.
2. *Modeling of workflows and guidelines.* Using the modeling language defined in step 1, models of workflows or specific treatment protocols are created. These models are formal representations that drive the management of clinical activities and services. The precise semantic foundation of the MIC modeling infrastructure and related tools enable validation and verification of the models against a range of safety-, privacy-, and security-related criteria defined as constraints or policies.
3. *Generation of process management systems.* Using the MIC model transformation infrastructure, the verified models are translated into configuration files that customize the generic run-time components (such as workflow and execution engines, graphical user interface, and interface to other components of the clinical environment) of the process management system [49].

Application of model-based methods in the design and operation of HISs is in its infancy. Combined with policy modeling, this approach has the promise of yielding a comprehensive approach to engineer privacy-aware HISs. In addition, application of formally represented guideline models can evolve with accumulated experience and can be customized to the needs of individual patients.

5.6 Model-Based Experiment Integration

Development of advanced HISs is an inherently experimental process. They require in-depth evaluation in the clinical environment to understand impact on quality of care. Unfortunately, creating credible experiments in the form of carefully controlled clinical trials is a complex task because of the need for integrating the experimental system with the usually heterogeneous health IT infrastructure, such as electronic medical records, order management systems, authentication systems, audit logs, whiteboards, and many others. Integration in real life environments can be prohibitively expensive without a systematic approach that helps rapid system integration.

One of the possible integration approaches is based on the use of the SOA framework. SOA provides an intuitive approach to resolve system diversity and integrate HISs. Instead of relying on site-specific, ad hoc design strategies, SOA provides a mechanism to create complex distributed applications from loosely coupled services. Workflow definitions define the structure of the applications, capturing the interaction of services with formally provable properties [15–17]. SOA inherently supports modular design, promotes reusability, and simplifies system evolution since services can be individually modified, upgraded, or completely replaced [50, 51]. SOAs have been successfully applied in the e-commerce sector [18] and, as a result, a rich infrastructure of SOA-specific tools is available [19, 20]. An essential advantage of the SOA framework is the extensive work on SOA security standards in the OASIS (Organization for the Advancement of Structured Information Standards) consortium that includes the Healthcare Information Technology Standards Panel (http://www.oasis-open.org/committees/xspa/charter.php#item-2).

There are several examples for applying SOA as foundation for HISs. Jürjens and Rumm provided an excellent summary of SOA and model-based approaches in clinical information systems [52]. Ruth Breu and collaborators at University of Innsbruck have extensive work on establishing an SOA-based security architecture, SECTET for HISs [53, 54]. Researchers at Vanderbilt have developed a Model-Integrated Clinical Information System architecture [55–57] that allows the use of simplified, domain-specific modeling languages to describe workflows and interfaces in restricted domains.

It is expected that integration technologies will fill in enabling role not only in experimentation, but also in the development and deployment of HISs.

5.7 Closing the Loop: Patient-Centered Health Information Management Evaluation

The ultimate goal of TRUST research on HIS is contribution to the vision of patient centered health information management. Despite their potential benefits, the national-scale deployment of HISs presents unique challenges, not unlike those in all other areas of critical infrastructure. The electronic nature of HISs, for instance, in contrast to the traditional paper-based systems, increases the potential magnitude of system failures and inadvertent exposure of highly sensitive patient-specific health information. It is necessary that HCOs design HISs to adhere to diverse governmental regulations that influence both procedural and access policies, such as the Privacy and Security Rules of the Health Insurance Portability and Accountability Act (HIPAA) [13, 14]. HIS must account for the complexities of health care delivery, both at the social and the technical level, to ensure timely access to health information and services without compromising the security of the system or sensitive patient records.

Patient centered health information systems advance toward increasingly patient-specific treatment strategies. These advancements including the related research and clinical practice occurs in the context of a regulatory environment that is increasingly privacy aware. Resolving the dichotomy between the goals of individualized medicine and preserving patient privacy is a major technological challenge.

References

[1] Kaushal, A., et al., "Return on Investment for a Computerized Physician Order Entry System," *J. Am. Med. Inform. Assoc.*, Vol. 13, No. 3, 2006, pp. 261–266.

[2] Shabo, A., "A Global Socio-Economic-Medico-Legal Model for the Sustainability of Longitudinal Electronic Health Records," *Methods Inf. Med.*, Vol. 45, No. 3, 2006, pp. 240–245.

[3] Simon, S. J., and S. J. Simon, "An Examination of the Financial Feasibility of Electronic Medical Records (EMRs): A Case Study of Tangible and Intangible Benefits," *Int. J. Med. Inform.*, Vol. 2, No. 2, 2006, pp. 185–200.

[4] Menachemi, N., and R. G. Brooks, "Reviewing the Benefits and Costs of Electronic Health Records and Associated Patient Safety Technologies," *J. Med. Syst.*, Vol. 30, No. 3, 2006, pp. 159–168.

[5] Lo, H. G., et al., "Electronic Health Records in Specialty Care: A Time-Motion Study," *J. Am. Med. Inform. Assoc.*, Vol. 14, No. 5, 2007, pp. 609–615.

[6] Pizziferri, L., et al., "Primary Care Physician Time Utilization Before and After Implementation of an Electronic Health Record: A Time-Motion Study," *J. Biomed. Inform.*, Vol. 38, No. 3, 2005, pp. 176–88.

[7] Silver, M. R., and R. Lusk, "Patient Safety: A Tale of Two Systems," *Qual. Manag. Health Care*, Vol. 10, No. 2, 2002, pp. 12–22.

[8] Frank, L., et al., "Using BPI and Emerging Technology to Improve Patient Safety," *J. Healthc. Inf. Manag.*, Vol. 18, No. 1, 2004, pp. 65–71.

[9] Shepherd, M., D. Zitner, and C. Watters, "Medical Portals: Web-Based Access to Medical Information," *Proceedings of the 33rd Annual Hawaii International Conference on System Sciences (HICSS-33)*, Maui, HI, January 4–7, 2000.

[10] Barnett, G. O., et al., "Overcoming Information Overload: An Information System for the Primary Care Physician," *Proceedings of the 11th World Congress on Medical Informatics (Medinfo 2004)*, San Francisco, CA, September 7–11, 2004.

[11] Masys, D., et al., "Giving Patients Access to Their Medical Records: The PCASSO Experience," *J. Am. Med. Inform. Assoc.*, Vol. 9, 2002, pp. 181–191.

[12] Cimino, J. J., V. L. Patel, and A. W. Kushniruk, "The Patient Clinical Information System (PatHIS): Technical Solutions for and Experience with Giving Patients Access to Their Electronic Medical Records," *Int. J. Med. Inform.*, Vol. 68, 2002, pp. 113–127.

[13] U.S. Department of Health and Human Services. "Standards for Privacy of Individually Identifiable Health Information, Final Rule," *Federal Register*, August 12, 2002, 45 CFR: Parts 160–164.

[14] U.S. Department of Health and Human Services, Office for Civil Rights, "Standards for Protection of Electronic Health Information, Final Rule," *Federal Register*, February 20, 2003, 45 CFR: Part 164.

[15] Yanchuk, A., A. Ivanyukovich, and M. Marchese, "Towards a Mathematical Foundation for Service-Oriented Applications Design," July 21, 2008, http://www.science.unitn.it/~marchese/pdf/Towards_SOAD_JoS_06.pdf.

[16] Portier B., "SOA Terminology Overview, Part 1: Service, Architecture, Governance, and Business Terms," July 21, 2008, http://www-128.ibm.com/developerworks/library/ws-soa-term1/index.html.

[17] Portier B. "SOA Terminology Overview, Part 2: Development Processes, Models, and Assets," July 21, 2008, http://www-128.ibm.com/developerworks/library/ws-soa-term2/index.html.

[18] Gray, J., "A Conversation with Werner Vogels, CTO, Amazon.com," *Web Services*, 2006, http://portal.acm.org/ft_gateway.cfm?id=1142065&type=pdf.

[19] Service-Oriented Architecture, http://www.oracle.com/technologies/soa/index.html.

[20] Service-Oriented Architecture (SOA), http://www.sun.com/products/soa/index.jsp.

[21] SOA Software – Solutions – SOA Fabric, http://www.soa.com/index.php/section/solutions/soa_fabric/.

[22] http://www.truststc.org.

[23] Department of Health and Human Services. "Standards for Privacy of Individually Identifiable Health Information: Final Rule," *Federal Register*, August 12, 2002, 45 CFR: Parts 160–164.

[24] U.S. Department of Health and Human Services, Office for Civil Rights, "Standards for Protection of Electronic Health Information: Final Rule," *Federal Register*, February 20, 2003; 45 CFR: Part 164.

[25] Masys, D. R., and D. B. Baker, "Patient-Centered Access to Secure Systems Online (PCASSO): A Secure Approach to Clinical Data Access Via the World Wide Web,"

Journal of the American Medical Informatics Association, Vol. 4, Fall Symp. Supplement, October 1997, pp. 340–343.

[26] Masys, D. R., et al., "PCASSO: A Secure Architecture for Access to Clinical Data Via the Internet," *Proceedings of the World Congress on Medical Informatics (MEDINFO '98), Intl. Med. Informatics Assoc.,* August 1998.

[27] Masys, D. R., et al., "Giving Patients Access to Their Medical Records: The PCASSO Experience," *Journal of the American Medical Informatics Association,* March/April 2002, pp. 181–191.

[28] Rehm, S., and S. Kraft, "Electronic Medical Records: The FPM Vendor Survey," *Family Practice Management,* Vol. 8, No. 1, 2001, pp. 45–54.

[29] Werner, J., et al., "Platform-Based Design for Clinical Information Systems," *Proceedings of the 5th IEEE International Conference on Industrial Informatics (INDIN),* Vienna, Austria, June 23–27, 2007, pp. 749–755.

[30] Duncavage, S., et al., "A Modeling Environment for Patient Portals," *Proceedings of the 2007 American Medical Informatics Association Annual Symposium (AMIA),* Washington, D.C., November 10–14, 2007, pp. 201–206.

[31] Barth, A., et al., "Privacy and Contextual Integrity: Framework and Applications," *Proceedings of 27th IEEE Symposium on Security and Privacy,* May 2006.

[32] Mathe, J., et al., "A Model Integrated Guideline-Driven Clinical Decision Support System," *IEEE Software,* Vol. 26, No. 4, July 2009, pp. 54–61.

[33] Oracle, "Best Practices for California SB 1386," http://www.oracle.com/technology/deploy/security/db_security/pdf/oracle_sb_1386.pdf, 2006.

[34] Li, N., and J. C. Mitchell. "Datalog with Constraints: A Foundation for Trust Management Languages," *Proceedings of the Fifth International Symposium on Practical Aspects of Declarative Languages (PADL 2003),* January 2003.

[35] Li, N., and J. C. Mitchell, "RT: A Role-Based Trust-Management Framework," *Proceedings of The Third DARPA Information Survivability Conference and Exposition (DISCEX III),* 2003, pp. 201–212.

[36] Li, N., J. C. Mitchell, and W. H. Winsborough, "Beyond Proof-of-Compliance: Security Analysis in Trust Management," *J. ACM,* Vol. 52, No. 3, 2005, pp. 474–514.

[37] Li, N., W. H. Winsborough, and J. C. Mitchell, "Distributed Credential Chain Discovery in Trust Management," *Journal of Computer Security,* Vol. 11, No. 1, February 2003, pp. 35–86.

[38] Blaze, M., J. Feigenbaum, and J. Lacy, "Decentralized Trust Management," *Proceedings of the 1996 IEEE Symposium on Security and Privacy,* May 1996, pp. 164–173.

[39] Ceri, S., G. Gottlob, and L. Tanca, "What You Always Wanted to Know About Datalog (and Never Dared to Ask)," *IEEE Trans. on Knowledge and Data Engineering,* Vol. 1, No. 1, 1989, pp. 146–166.

[40] Jackson, E., and J. Sztipanovits, "Formalizing the Structural Semantics of Domain-Specific Modeling Languages," *Journal of Software and Systems Modeling,* September 2009, pp. 451–478.

[41] Chen, Y., and B. Malin, "Detection of Anomalous Insiders in Collaborative Environments Via Relational Analysis of Access Logs," *Proceedings of the ACM Conference on Data and Application Security and Privacy,* San Antonio, TX, 2011 (in press).

[42] Bate, D. W., "The Quality Case for Information Technology in Healthcare," *BMC Medical Informatics and Decision Making,* 2002, 2:7 doi:10.1186/1472-6947-2-7.

[43] Goldman, J., and Z. Hudson, "Virtually Exposed: Privacy and E-Health," *Health Affairs,* November/December 2000, http://content.healthaffairs.org/cgi/search?ck=nck.

[44] The Markle Foundation, Connecting for Health: "A Common Framework for Networked Health Information," December 7, 2006, http://www.connectingforhealth.org/commonframework/docs/P9_NetworkedPHRs.pdf.

[45] Ledeczi, A., et al., "Composing Domain-Specific Design Environments," *Computer,* November 2001, pp. 44–51.

[46] Balasubramanian, K., et al., "Developing Applications Using Model-Driven Design Environments," *IEEE Computer,* February 2006, pp. 33–40.

[47] Karsai, G., et al., "The Model Integrated Computing Toolsuite: Metaprogrammable Tools for Embedded Control System Design," *Proceedings of the IEEE Joint Conference CCA, ISIC and CACSD,* Munich, Germany, 2006.

[48] Emerson, M., J. Sztipanovits, and T. Bapty, "A MOF-Based Metamodeling Environment," *Journal of Universal Computer Science,* Vol. 10, No. 10, October 2004, pp. 1357–1382.

[49] The Generic Modeling Environment, http://www.isis.vanderbilt.edu/projects/gme/.

[50] Portier, B., "SOA Terminology Overview, Part 1: Service, Architecture, Governance, and Business Terms," http://www-128.ibm.com/developerworks/library/ws-soa-term1/index.html.

[51] Portier, B., "SOA Terminology Overview, Part 2: Development Processes, Models, and Assets," http://www-128.ibm.com/developerworks/library/ws-soa-term2/index.html.

[52] Jürjens, J., and R. Rumm, "Model-Based Security Analysis of the German Health Card Architecture," *Methods of Information in Medicine,* Vol. 47, No. 5, 2008, pp. 26–35.

[53] Alam, M., M. Hafner, and R. Breu, "A Constraint Based Role Based Access Control in the SECTET: A Model-Driven Approach," *Journal of Computer Security,* Vol. 16, No. 2, 2008, pp. 223–260.

[54] Alam, M., R. Breu, and M. Hafner,"Modeling Permissions in a (U/X)ML World," *First International Conference on Availability, Reliability and Security (ARES'06),* Vienna, Austria, April 20–22, 2006, pp. 685–692.

[55] Mathe, J., et al., "Implementing a Model-Based Design Environment for Clinical Information Systems," *MOTHIS 2007,* Nashville, TN, September 30, 2007.

[56] Werner, J., et al., "Platform-Based Design for Clinical Information Systems," *Proceedings of the 5th IEEE International Conference on Industrial Informatics (INDIN),* Vienna, Austria, 2007, pp. 749–755.

[57] Mathe, J., et al., "Model-Based Design of Clinical Information Systems," *Methods of Information in Medicine,* Vol. 47, No. 5, 2008, pp. 10–21.

6

Urban Defense Using Mobile Sensor Platforms: Surveillance, Protection, and Privacy

Mario Gerla, Uichin Lee, Nanxi Chen, and Soon Y. Oh

6.1 Introduction

Video cameras are proliferating at very fast pace in urban areas all over the world. The video is now becoming the most effective way to apprehend criminals and, in some cases, to prevent crimes. In this chapter we make the case for complementing the traditional video cameras installed in the infrastructure (walls, light poles, bridges) with mobile video cameras mounted on vehicles. This opportunity opens a whole new gamut of possibilities. It also poses questions about privacy and security, as well as social acceptance. In this chapter we address these issues and open new questions for future researchers in the field.

6.1.1 Urban Surveillance

The visitors of Palazzo dei Dogi (Dogi' Palace) in Venice learn that the security of the Venice Republic and its citizens against internal plots or foreign intrusions was assured by two important instruments: government spies and anonymous informants. The latter would use the ill-famed bocca della verita' (the mouth of truth) to drop their notes. Today homeland defense against attacks of all sorts relies heavily on the new tools of the modern technology. Massive computers and databases can track suspects in unimaginable ways. However, the spy and informant instruments still exist, albeit in a new, more efficient form enabled by the technology. The spies have been replaced by video surveillance—we see video cameras popping up everywhere in the city, from the traffic intersection, to the subway platforms and to your own "private" hotel room. The informants today are equipped with smart phones and upload personal blogs and YouTube video clips (in some cases, anonymously) to popular Web sites. In fact, the uploading of blogs and "tweets" from smart phones is not just a form of entertainment and social networking for teenagers; it is becoming an important element for first responders' assistance and emergency recovery. Firefighters can use these video clips to judge the criticality of fire/flood situation and can do effective triage based on such reports.

In this chapter we focus on video surveillance as opposed to active (i.e. human controlled) surveillance using handheld phones with cameras. Video surveillance is still one of the major inputs to homeland security today, as we know from recent events in Times Square, New York, where the terrorist was caught on camera while setting up the bomb and fleeing the scene. Had the bomb actually detonated (which fortunately did not), the video camera would have been the only witness. This observation has a corollary. If we had been able to follow the perpetrator on his way back to the car, we could have been able to read his license number and eventually apprehend him even if the bomb had exploded obliterating all traceable identifiers. However, the wall cameras are not sufficient to track pedestrians or cars step by step as they move around in a large urban area. Enter the concept of mobile cameras placed on cars. Future cars will see and record everything within view (this costs nothing and will probably be recommended by insurance companies). So, passing cars could track the terrorist on his way back to the car.

This example helps to introduce the message that this chapter wants to convey: use mobile video surveillance platforms to complement the existing fixed installations. In the rest of this section, we review the state of the art of video surveillance in major cities in the world, report recent trends, and then comment on the limitations of infrastructure mounted video cameras.

More generally, in this chapter we examine vehicular platforms; propose efficient vehicular protocols and applications for surveillance; address the unique security and privacy requirements posed by homeland security applications; and finally describe a testbed, currently under development at UCLA, that will be used to test and demonstrate these concepts.

6.1.2 Video Camera Surveillance

Video surveillance camera networks have been invading all major cities in the world. There exist about 25 million closed-circuit television (CCTV) cameras worldwide (see Figure 6.1). They watch passengers in the street and stations, monitor cashiers at convenience stores, and capture identities of customers at restaurants. They can tilt, pan, and zoom, and are located on building roofs or lampposts in urban districts. Furthermore, some of them are able to predict people's behavior and motion in order to recognize possible dangerous situations.

The United Kingdom is well known for featuring the largest per capita deployment, and London has more surveillance cameras monitoring citizens than any other cities in the world. In the 1970s, the U.K. police began installing cameras at sport stadiums to track and stop violence at soccer games. After a bombing attack by the Irish Republican Army in 1993, authorities moved on to install cameras in the financial district in London, the often criticized Ring of Steel. Lots of private cameras in London offices, banks, and businesses have been incorporated into the police monitoring system, called CameraWatch [1]. Experts estimate that 4.2 million CCTVs (the CCTV User Group insists 7 million CCTVs [2]) are currently in operation in the United Kingdom, which is one for every 14 residents, and 500,000 CCTV cameras are watching citizens at car parking lots, bus stops, and train stations in London alone. Thus, anyone living or working in London is likely to be captured on camera on average 300 times a day. For this system, the U.K. police spend two-thirds of its budget to install and maintain cameras every year as a measure for deterring crime.

Among big cities in the United States, Chicago is known to have the most extensive surveillance camera network. By spending millions, Chicago was targeting one camera on every corner of the urban district in preparation for their 2016 Summer Olympic Games bid (which was not granted). Experts estimate that 15,000 cameras are connected in the Chicago surveillance system, which is called Operation Virtual Shield [3]. Chicago allows private companies to join the surveillance system deployment. For example, with IBM it is planning to develop the most advanced video security network based on object recognition technology. A license plate recognition program inside the camera sends an alert message to the emergency center once a camera detects a suspect license

Figure 6.1 Surveillance cameras and a monitoring center.

number. As another application, the programmed camera detects a suitcase left unguarded in a public place. They have shown that the automated detection system is more effective (and less costly) than employees screening the monitored video. IBM has also been working with the Chicago Transit Authority to install real-time surveillance system on the Chicago public buses. Real-time video, which records the interior of the bus, is streamed to the control center via wireless channels through access points at the bus stops. This way, police headquarters and first responders can continuously monitor the situation and be ready for intervention. Furthermore, a video camera mounted in front of the bus collects and stores video that can be used to resolve disputes in case of accidents.

Other cities in the United States also have plenty of surveillance cameras, but their systems are not as well organized and managed as the Chicago system. New York spent $90 million for its monitoring network by 4,000 cameras in Lower Manhattan after 9/11 and has announced $24 million expansion of the network to cover midtown, including Times Square [5] (see Figure 6.2). However, it is known that about half of the 4,300 cameras installed along the New York subways are not operational. Moreover, other cities in the United States (e.g., Baltimore and Washington, D.C.) have installed lots of surveillance cameras, but they have not yet integrated surveillance systems with the city's emergency system [6] (see Figure 6.3).

The surveillance system is rapidly spreading because of citizens' increasing concern about crime and terrorism. Indeed, public polls largely support the use of such surveillance camera systems. For instance, a 2006 Harris Poll found that two-thirds of Americans agree with the expansion of video surveillance [5]. In addition, police and city officers stress that camera systems are critical

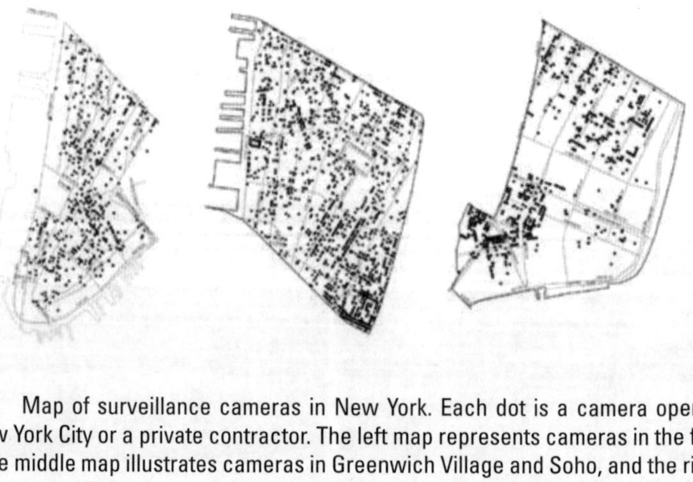

Figure 6.2 Map of surveillance cameras in New York. Each dot is a camera operated by either New York City or a private contractor. The left map represents cameras in the financial district, the middle map illustrates cameras in Greenwich Village and Soho, and the right map describes cameras in the Lower East Side and Chinatown. (From: [4].)

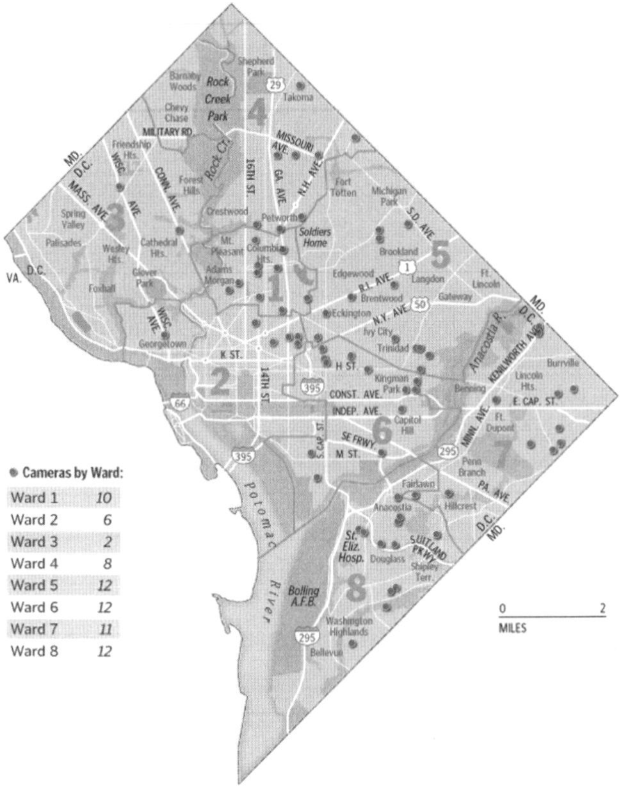

Figure 6.3 Map of surveillance cameras in Washington, D.C., District 73. (From: [7].)

to prevent crimes and also to prosecute criminals. Chicago police asserts that 4,000 arrests were made since 2006 and 30% crime reduction was achieved with the help of cameras. The British police also credit cameras for helping to solve major crimes by tracking the movement of individuals suspected to be involved in criminal activities. The increased popularity of video surveillance has stimulated the development of more advanced video technology. For instance, under funding of the U.S. Department of Homeland Security's Science and Technology Directorate (S&T) a new video camera called Imaging System for Immersive Surveillance (ISIS) was developed, which can record 360° video with high quality [8].

There are still, however, many concerns and counterarguments against video surveillance. Many experts and civil rights organizations insist that CCTV cameras cannot effectively prevent crimes and cannot significantly decrease criminal rate. The British Home Office has studied the impact of surveillance cameras, but was not able to find any positive correlation with criminal rate reductions. Furthermore, the study finds that people do not feel safer in the presence of cameras. In fact, the London surveillance system was unable to deter the bombing attacks in July 2005 and June 2007. Moreover, a 2008 report by the U.K. police chief concluded that only 3% of crimes are solved by cameras. Due to this evidence, experts and organizations say that the system is only useful after the crime has happen and it simply causes a displacement effect—crimes move away to areas where there are no cameras. Thus they further argue that the police must stop spending taxpayer money to expand the number of surveillance cameras. Another serious concern is privacy abuse. A person starts to be recorded once he/she steps out the house and his personal information is available to anyone that has access (legally or not) to the surveillance system. Truly, this happened in 2004 in New York when a couple on a date on the rooftop of their building was secretly watched by police officers charged with conducting surveillance of protest rallies during the Republican National Convention [4].

Like opponents argue, the fixed CCTV surveillance system has several disadvantages. Some include:

- It is expensive to install and to maintain CCTVs.
- There are still many blind areas in the city.
- It is difficult to find and trace an event from tremendous amount of video contents.
- Privacy is critical.

Recently, mobile sensors (e.g., vehicular and smart-phone sensors) provide tremendous opportunities to situation monitoring and distributed surveillance so that they can be an excellent complement to the fixed surveillance system. Mobile sensors are able to cover most areas of the city so that they make it harder for potential attackers to disable surveillance. In addition, they can use proactive monitoring and filtering where vehicles continuously sense events from urban streets, maintain sensed data in their local storage, autonomously process them (e.g., recognizing license plates), and possibly route messages to vehicles in their vicinity to achieve a common goal (e.g., to permit police agents to track the movements of specified cars). In this chapter, we present an urban monitoring platform on vehicular and address security and privacy requirements in surveillance applications relevant to homeland security.

6.1.3 Fixed Video Camera Limitations and Emergence of Mobile Sensing

Video cameras are proliferating at very fast pace. They offer very important benefits. First, the CCTV technology (possibly supported by fibers) permits to collect economically, in real time, high resolution video at central control stations. Second, advances in processing permit online, real time processing of the video, and identify threats automatically, without relying on visual screening. Face recognition may soon become a reality. However, fixed video cameras have some well known drawbacks. They are still relatively expensive and thus cannot be installed with sufficient density to actually track pedestrians. They can be taken out by a determined terrorist or, more likely, by hostile citizens who resent being observed. The central facilities can be sabotaged. Finally, the Panopticon effect (i.e., the feeling of being observed by cameras everywhere) makes even the most cooperative citizens uncomfortable. Our vision is that these problems can be overcome by complementing the fixed cameras with pervasive cameras installed on citizens' vehicles.

As mentioned earlier, mobile sensor platforms such as personal smart phones and vehicular radios are gradually emerging as critical components of an urban homeland security program. Vehicles will complement the fixed surveillance infrastructure by providing a ubiquitous, mobile sensing (acoustic, video, chemical, radiation) capability. Vehicles can monitor the environment for potential threats and for after the fact forensic investigation of attacks (for instance, tracing the approach route of a suicide bomber to find accomplices). Likewise, personal phones can be used for a pervasive urban sensing at a level not accessible via fixed sensors embedded in the infrastructure. Personal phones can be used as passive sensors, for example, to monitoring representative samples of the population for possible bio-attacks. They can also be used under direct owner control to upload relevant and very valuable video clips to the Internet, as recent events have amply demonstrated. Personal phones, however, have limited battery life and also require in some cases active participation by the owner. These facts limit the use of phones to very specific applications.

Interestingly, these mobile platforms also represent valuable communication backups in the case of urban infrastructure massive failure due to natural disaster or hostile attack. After the attack, vehicles and phones can self-organize to form a very robust backup network fabric that, jointly with surviving wireless mesh nodes (e.g., solar powered traffic lights) can guarantee first responder communications and coordination.

In Section 6.2 we introduce a proposed architecture for mobile sensing: MobEyes. We discuss the challenges of storing the video on vehicles, of maintaining a distributed index to such video files, and of harvesting the files of interest using mobile agents. We then demonstrate, in Section 6.3 the efficiency of vehicular surveillance in two representative applications.

Critical issues when private citizens' platforms are involved are privacy (i.e., anonymous, untraceable sources) and security (selective authorization to receive sensitive information on an as needed basis). We address these issues in Section 6.4.

6.2 MobEyes: Video Surveillance from Vehicular Platforms

Recent advances in vehicular communications make it possible to realize vehicular sensor networks, that is, collaborative environments where mobile vehicles equipped with sensors of different natures (from toxic detectors to still/video cameras) interwork to implement monitoring applications. Namely, vehicles continuously sense events from urban streets, autonomously process sensed data (e.g., recognizing license plates), and exchange messages with neighbors to achieve a common goal, say, to permit police agents to track the movements of specified cars. This challenging environment requires novel solutions, with respect to those of more traditional wireless sensor nodes, like fixed video cameras. In fact, vehicles exhibit constrained mobility, have no strict limits on processing power and storage capabilities, and host sensors that may generate sheer amounts of data, thus making inapplicable already known solutions for sensor network data reporting.

This section describes MobEyes, an effective middleware specifically designed for proactive urban surveillance exploiting node mobility to opportunistically diffusion of data summaries. This section includes: evaluation of MobEyes performance as a function of vehicle mobility, effects of concurrent exploitation of multiple harvesting agents with single/multihop communications, and MobEyes validation in a challenging urban tracking application.

6.2.1 Vehicular Sensor Networks

Vehicular ad hoc networks (VANETs) are acquiring commercial relevance because of recent advances in intervehicular communications and decreasing costs of related equipment. In particular, vehicular sensor networks (VSN) are emerging as a new tool for effectively monitoring the physical world, especially in urban areas where a high concentration of vehicles equipped with on board sensors (GPS, chemical spill detectors, still/video cameras, vibration sensors, acoustic detectors) is expected [9]. Vehicles are typically not affected by strict energy constraints and can be easily equipped with powerful processing units, wireless transmitters, and sensing devices even of some complexity, cost, and weight. VSN represent a significantly novel and challenging deployment scenario, considerably different from more traditional wireless sensor network environments, thus requiring innovative specific solutions. In fact, different from wireless sensor nodes, vehicles usually exhibit constrained mobility patterns due to street layouts, junctions, and speed limitations. In addition, they usually have no strict limits on processing power and storage capabilities. Most important, they can host sensors that may generate huge amounts of data, such as multimedia video streams, thus making impractical the instantaneous data reporting solutions of conventional wireless sensor networks.

VSN offer a tremendous opportunity for a broad range of large-scale applications, from intelligent transport (i.e., traffic routing and congestion mitigation) to environmental monitoring and distributed surveillance. In particular, there is an increasing interest in proactive urban monitoring services where vehicles continuously sense events from urban streets, maintain sensed data in their local storage, autonomously process them (e.g., recognizing license plates), and possibly route messages to vehicles in their vicinity to achieve a common goal (e.g., to permit police agents to track the movements of specified cars). For instance, proactive urban monitoring could usefully apply to post facto crime scene investigation. Reflecting on tragedies such as September 11, 2001, and the London bombings, VSN could have actually helped emergency recovery and forensic investigation/criminal apprehension. In the London bombings police agents were able to track some of the suspects in the subway using closed-circuit TV cameras, but they had hard time finding helpful evidence from the double-decker bus; this has motivated the installation of more cameras in fixed locations along London streets. VSN could be an excellent complement to the deployment of fixed cameras/sensors. The completely distributed and opportunistic cooperation among sensor-equipped vehicles has the deterrent effect of making it harder for potential attackers to disable surveillance. Another less sensational but relevant example is the need to track the movements of a car, used for a bank robbery, in order to identify the perpetrators, say. It is highly probable that some vehicles have spotted the unusual behavior of the suspects' car in the hours before the robbery, and might be able to identify the threat by opportunistic peer to peer (P2P) more difficult for the police to extract that information from the massive number of multimedia streams recorded by fixed cameras. As for privacy, let us briefly note that people are willing to sacrifice privacy and to accept a reasonable level of surveillance when the data can be collected and processed only by recognized authorities (with a court order), for a valid purpose (e.g., forensic crime investigation and/or terrorism prevention).

As shown by the above examples, the reconstruction of a crime and, more generally, the forensic investigation of an event monitored by VSN require the collection, storage, and retrieval of massive amounts of sensed data. This is a major departure from conventional sensor network operations where data is dispatched to sinks under predefined conditions such as alarm thresholds. Obviously, it is impossible to deliver all the streaming data collected by video sensors to a police authority sink because of sheer volume. Moreover, input filtering is not possible because a priori nobody knows which data will be of use for future investigations. To that purpose, we have designed and implemented MobEyes, a novel middleware that supports VSN-based proactive urban monitoring applications. MobEyes exploits wireless-enabled vehicles equipped with video cameras and a variety of sensors to perform event sensing, processing/classification of sensed data, and intervehicle ad hoc message routing. Since it is impossible to directly report the sheer amount of sensed data to the authority, MobEyes keeps sensed data in mobile node storage; on board processing capabilities are used to extract features of interest (e.g., license plates); mobile nodes periodically generate data summaries with extracted features and context information such as timestamps and positioning coordinates; and mobile agents (e.g., police patrolling cars) move and opportunistically harvest summaries as needed from neighbor vehicles. MobEyes adopts VSN custom designed protocols for summary diffusion/harvesting that exploit intrinsic vehicle mobility and simple single-hop intervehicle communications. In that way, MobEyes harvesting agents can create a low-cost opportunistic index to query the distributed sensed data storage, thus enabling us to answer questions such as: which vehicles were in a given place at a given time, which route did a certain vehicle take in a given time interval, and which vehicle collected and stored the data of interest?

6.2.2 Related Work

6.2.2.1 Vehicle Sensor Platforms

The idea of embedding sensors in vehicles is very novel. To our knowledge, the only well publicized research project dealing with extensive vehicle sensor data mining is MIT's CarTel [10, 11]. In CarTel users submit their queries about sensed data on a portal hosted on the wired Internet. Then, an intermittently connected database is in charge of dispatching queries to vehicles, and of receiving replies when vehicles move in the proximity of open access points to the Internet. Other interesting research projects have focused on the close area of providing mobile Internet access to vehicles. For instance, InternetCar aims to provide vehicles with seamless Internet connectivity, by envisioning various applications such as a traffic information dissemination service where raw sensed data from vehicles are collected in a central server and traffic information is distributed to the drivers [12, 13]. Differently from CarTel and InternetCar, MobEyes exploits mobile collector agents instead of relying on the wired Internet infrastructure, thus improving robustness. Note that also the FleetNet project, which aims to develop an intervehicle communication platform for vehicular applications, recognized the potential of services that distribute location-tagged information (e.g., traffic jam warning) by collecting and processing data from cars in a distributed fashion [14, 15]. In this paper, we originally propose how to design and implement such applications using vehicular sensor networks.

6.2.2.2 Opportunistic Sensor Networking

Traditionally, sensor networks have been deployed in static environments, with application-specific monitoring tasks. Recently, opportunistic sensor networks have emerged, which exploit existing devices and sensors, such as cameras in mobile phones [16–19]. Several of these networks are relevant to our research because they can easily implement opportunistic dissemination protocols [20, 21].

Dartmouth's MetroSense [16, 22] is closely related to MobEyes. Reference [16] describes a three tier architecture for MetroSense: servers in the wired Internet are in charge of storing/processing sensed data; Internet-connected stationary sensor access points (SAP) act as gateways between servers and mobile sensors (MS); MS move in the field opportunistically delegating tasks to each other, and muling [23, 24] data to SAP. MetroSense requires infrastructure support, including Internet-connected servers and remotely deployed SAP. Similarly, Wang et al. proposed data delivery schemes in Delay/Fault-Tolerant Mobile Sensor Network (DFT-MSN) for human-oriented pervasive information gathering [25]. The trade-off between data delivery ratio/delay and replication overhead is mainly investigated in the case of buffer and energy resource constraints. In contrast, MobEyes does not require any fixed infrastructure by using mobile sinks (or agents) and addresses VANET-specific deployment scenarios (e.g., powerful sensing platforms, distributed index collection).

Among recent research projects about opportunistic sensing, we mention Intel IrisNet [26] and Microsoft SenseWeb [27]. Both projects investigate the integration of heterogeneous sensing platforms in the Internet via a common data publishing architecture. Finally, we point out CENS' Urban Sensing project [19, 28], a recently started multidisciplinary project addressing participatory sensing, where urban monitoring applications receive data from mobile sensors operated by people.

As for dissemination of sensed data through peers, we can mention two solutions from the naturalistic environment, namely ZebraNet [20] and SWIM [21]. ZebraNet addresses remote wildlife tracking (e.g., zebras in Mpala Research Center in Kenya) by equipping animals with collars that embed wireless communication devices, GPS, and biometric sensors. As GPS-equipped animals drift within the park, their collars opportunistically exchange sensed data, which must make its way to the base station (the ranger's truck). ZebraNet proposes two dissemination protocols: a flooding-based approach where zebras exchange all the data within their buffers (either locally generated or received from other animals) with neighbors and a history-based protocol where data is uploaded only to zebras with good track record of base station encounters. SWIM [21] addresses sparse mobile sensor networks with fixed infostations as collecting points. Sensed data is epidemically disseminated via single-hop flooding to encountered nodes and offloaded when info stations are in reach.

While all the above schemes implement peer-to-peer dissemination, none fits the performance requirements of the MobEyes target scenario. Flooding generates excessive overhead, while history-based is ineffective in the very dynamic VANET where the base station (the agent's vehicle) rapidly moves without following a specific mobility pattern [20]. In addition, let us note that MobEyes nodes do not transmit raw collected data but, thanks to an abundance of on board computing resources, they locally process data and relay only short summaries. That is a relevant advantage since data collected in VSN (e.g., from video cameras) may be an order-of-magnitude larger than data sensed in naturalistic scenarios.

6.2.3 MobEyes Architecture

For the sake of clarity, we present the MobEyes solution using one of its possible practical application scenarios: collecting information from MobEyes-enabled vehicles about criminals that spread poisonous chemicals in a particular section of the city (say, subway station). We suspect that the criminals have used vehicles for the attack. Thus, MobEyes helps detect the vehicles, and permits tracking and capture. Here, we assume that vehicles are equipped with cameras and chemical detection sensors. Vehicles continuously generate a huge amount of sensed data, store it locally, and periodically produce short summary chunks obtained by processing sensed data (e.g., license plate numbers or aggregated chemical readings). Summary chunks are aggregated in summaries that are opportunistically disseminated to neighbor vehicles, thus enabling metadata harvesting by the police in order to create a distributed metadata index, useful for forensic purposes such as crime scene reconstruction and criminal tracking.

To support all the above tasks, we have developed MobEyes according to the component-based architecture depicted in Figure 6.4. The key component is the MobEyes Diffusion/Harvesting Processor (MDHP), which is discussed in detail in the next section. MDHP works by opportunistically disseminating/harvesting summaries produced by the MobEyes Data Processor (MDP), which accesses sensor data via the MobEyes Sensor Interface (MSI). Since vehicles are not strictly resource-constrained, our MobEyes prototype is built on top of the

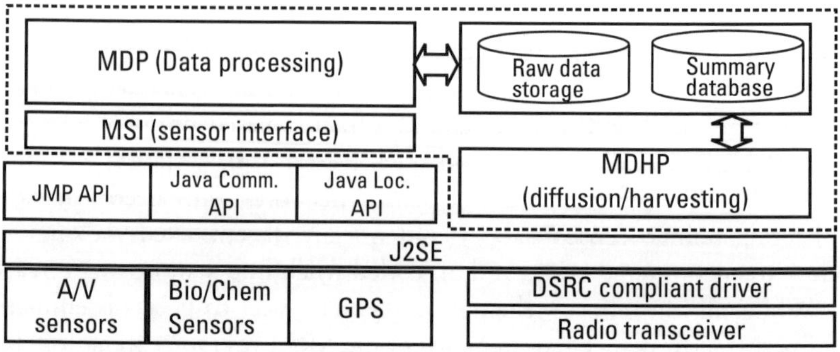

Figure 6.4 MobEyes sensor node architecture.

Java Standard Edition (J2SE) virtual machine. MDP is in charge of reading raw sensed data (via MSI), processing it, and generating chunks. Chunks include metadata (vehicle position, timestamp, vehicle ID number, and possible additional context such as simultaneous sensor alerts) and features of interest extracted by local filters (see Figure 6.5). For instance, in the above application scenario, MDP includes a filter determining license plate numbers from multimedia flows taken by cameras [29]. Finally, MDP commands the storage of both raw data and chunks in two local databases. MDHP disseminates/harvests summaries by packing a set of chunks into a single packet for the sake of effectiveness. Therefore, the generation rate and size of chunks and summaries are relevant to MobEyes performance. Additional details about the design and implementation of the MobEyes prototype are out of the scope of this paper and can be found in [30].

Developers of MobEyes-based applications can specify the desired generation rate as a function of vehicle speed and expected vehicle density. The chunk size mainly depends on application-specific requirements: in the scenario under consideration, each recognized license plate is represented with 6B, sensed data with 10B (e.g., concentrations of potential toxic agents), timestamp with 2B, and vehicle location with 5B. Then, in our scenario, MDP can pack 65 chunks in a single 1,500B summary, even without exploiting any data aggregation or encoding techniques. In usual deployment environments chunks are generated every 2–10 seconds and, thus, a single summary can include all the chunks about a 2–10-minute interval. MSI permits MDHP to access raw sensed data independently of actual sensor implementation, thus simplifying the integration with many different types of sensors. MSI currently implements methods to access camera streaming outputs, serial port I/O streams, and GPS information, by only specifying a high-level name for the target sensor. To interface

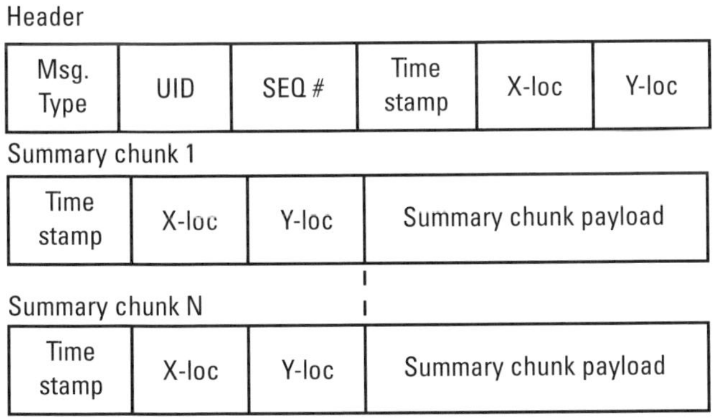

Figure 6.5 Packet format: a single packet contains multiple summary chunks corresponding to different events.

with sensor implementations, MSI exploits well known standard specifications to achieve high portability and openness: the Java Media Framework (JMF) API, the Sun Communication API, and the JSR179 Location API.

6.3 MobEyes Urban Surveillance Examples

6.3.1 The Westwood Scenario

In this section we evaluate the MobEyes surveillance efficacy in representative scenarios using the ns-2 simulator [31]. The first scenario tests the impact of vehicular motion (speed, traffic pattern) on sensor data collection. We consider vehicles moving in a fixed region of size 2,400m × 2,400m. The default mobility model is real-track (RT), introduced by our colleagues in [32]. RT permits to model realistic vehicle motion in urban environments. In RT, nodes move following virtual tracks, representing real accessible streets on an arbitrary loaded roadmap. For this set of experiments, we used a map of the Westwood area in the vicinity of the UCLA campus, as obtained by the U.S. Census Bureau data for street-level maps [33] (Figure 6.6). At any intersection, each node randomly selects the next track it will run through; speed is periodically allowed to change (increase or decrease) by a quantity uniformly distributed in the interval (0, ± Δs). To evaluate the impact of the mobility model on MobEyes performance, we tested two additional well-known models, namely Manhattan (MAN) [34] and Random WayPoint (RWP) [35]. Similarly to RT, MAN builds node trajectories following urban roads; however, in MAN roads are deployed according to a regular grid, thus allowing a more uniform node deployment. In our simulation, we adopted a 10 × 10 grid. RWP instead does not constrain node positions to follow actual road tracks, but moves nodes toward randomly selected destinations with random speeds. When a node reaches its destination, it pauses for a fixed period (which we set equal to 0 by homogeneity with the

Figure 6.6 Map of Westwood area in vicinity of UCLA campus.

other models), and then selects a new destination. Surprisingly, RWP is considered "a good approximation for simulating the motion of vehicles on a road" [36], generally producing limited distortion on protocol performance. Let us remark that MobEyes agents do not exploit any special trajectory or controlled mobility pattern, but move conforming to regular nodes.

Our simulations consider number of nodes $N = 100, 200, 300$. Vehicles move with average speed $v = 1.4, 4.1, 7$ m/s. To obtain these values with the random way mobility model, we tuned the minimum speed to $v_m = 1$ m/s and the maximum speed to $v_M = 11, 31, 51$ m/s, respectively. The summary advertisement period of regular nodes and the harvesting request period are kept constant and equal to 3*s* through all the simulations. We use a Bloom filter with 8,192 bits (1,024B) and 10 hash functions. Since we have a large filter size (i.e., 8,192 bits) compared to the number of summaries, the false positive probability is negligible.[1] We note that if the value of this parameter is too large, MobEyes effectiveness is reduced since it is possible that two nodes do not exchange messages, even if they occasionally enter in each other transmission range; this effect is magnified, as node speed v increases. The chosen value has been experimentally determined to balance the effectiveness of our protocol and the message overhead, even in the worst case (i.e., $v = 7$ m/s). A deeper and more formal investigation of the optimal value of the advertisement period is the object of future work.

Finally, we modeled communications as follows: MAC protocol IEEE 802.11, transmission band 2.4 GHz, bandwidth 11 Mbps, nominal radio range equal to 250m, and two-ray ground propagation model [37]. The values of these parameters have been chosen similar to other work in the field [38, 39]. Where not differently stated, reported results are average values out of 35 repetitions. Other MobEyes configuration parameters are introduced in the following subsections, when discussing the related aspects of MobEyes performance.

6.3.2 Data Validation

We investigate the regular node collection and agent harvesting processes. Without loss of generality, let us assume that new summaries are synchronously generated by all regular nodes. A generation epoch is the time interval between two successive summary generations. In this set of experiments, every regular node continuously advertises the single summary it generated in the epoch $t = 0$ for the rest of the simulation run.

Figures 6.7 and 6.8 show results collected for number of nodes $N = 100/300$, average speed $v = 1.4/7$ m/s, and RWP, MAN, and RT mobility models. In particular, Figure 6.7 plots the cumulative distribution of summaries

1. The false positive probability is $p_f = (1-(1-1/\mathrm{m})^{lw})^l \approx (1-e^{lw}/m)^l$ where m is total number of bits, w is the number of elements, and l is the number of hash functions.

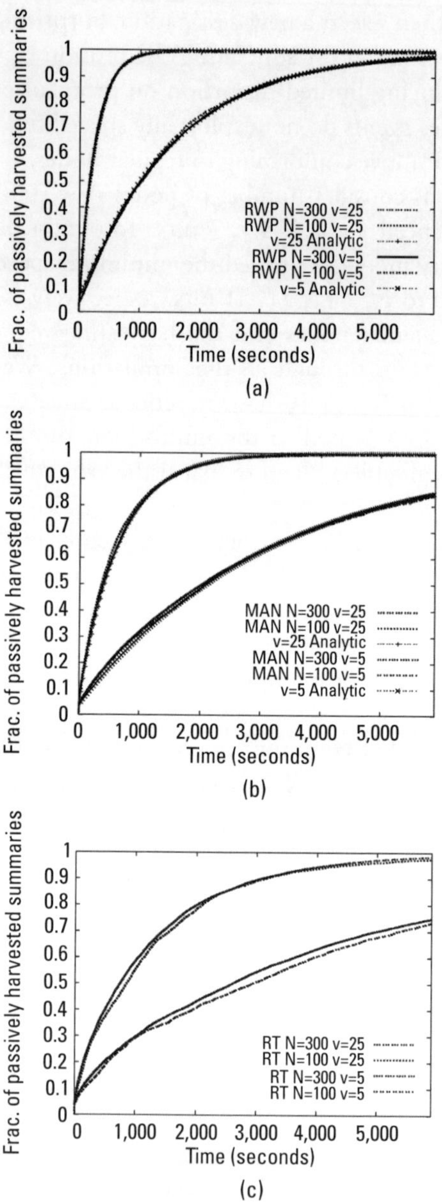

Figure 6.7 Fraction of passively harvested summaries by a regular node: (a) RWP, (b) MAN, and (c) RT.

collected by regular nodes as a function of time. The figure shows that the process highly depends on the average node speed; in fact, the speed determines to a large extent how quickly nodes infect other participants with their own summaries. The results do not depend on the number of nodes. The curves for the RT model exhibit worst fitting; they start deviating from the analysis

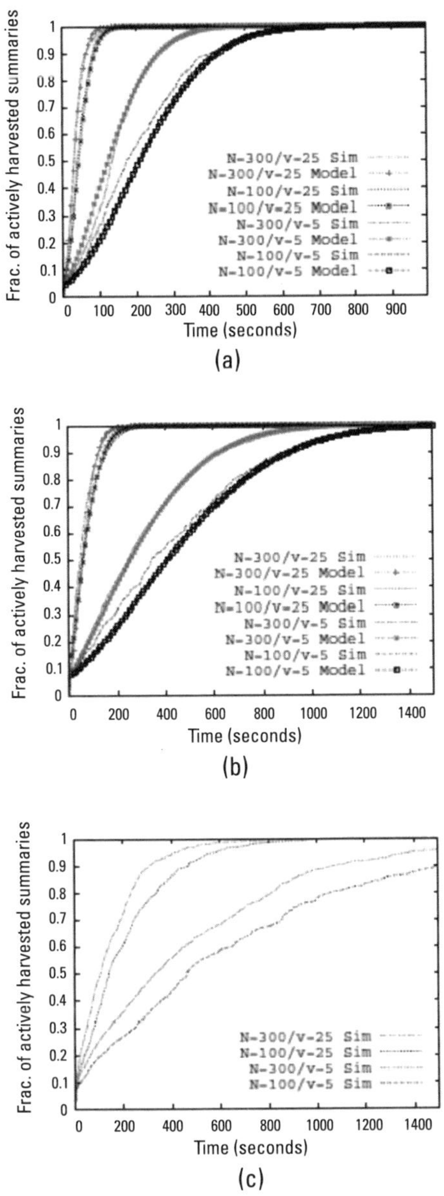

Figure 6.8 Fraction of actively harvested summaries by an agent: (a) RWP, (b) MAN, and (c) RT.

curves after certain thresholds. RWP shows slightly better accuracy mainly due to unconstrained motion and to the nodes' tendency to gather at the center of the field as time passes [40]. Although both MAN and RT show the restricted mobility patterns, their node distributions are different: MAN has almost uniform node distribution over the grids, whereas RT exhibits nonuniform node distribution over the map as shown in [41]. Since our model is based on uniform distribution, the curve fitting works well with MAN.

Figure 6.8 plots the cumulative distribution of summaries harvested by a police agent as a function of time. The figure shows that the completion times are mainly dependent on the speed. Unlike passive harvesting, the density plays an important role in active harvesting. In the analysis section, we show that the higher the density, the faster is the harvesting progress. Intuitively, if there are more neighbors, the agent has a higher chance of getting a random summary. Our analytic model fits well the simulations, especially when we have large N and v values.

6.3.3 Tracking Suspects Using MobEyes

For the sake of proving MobEyes effectiveness in supporting urban monitoring, we have also simulated a vehicle tracking application where the agent reconstructs node trajectories exploiting the collected summaries. This is a challenging application, since it requires our system: (1) to monitor a large number of targets (i.e., all participant vehicles); (2) to periodically generate fresh information on these targets, since they are highly mobile; and (3) to deliver to the agent a high share of the generated information. Moreover, since nodes are generally spread all over the area, this application shows that a single agent can maintain a consistent view of a large zone of responsibility. As regular cars move in the field, they generate new summaries every $T = 120$ seconds and continuously advertise the last generated summary. Every summary contains 60 summary chunks, which are created every *ChunkPeriod* = *2s* and include the license plate and position of the vehicle nearest to the summary sender at the generating time, tagged with a timestamp. The application exploits the MobEyes diffusion protocol where only the originator forwards the message (i.e., $k = 1$) to spread the summaries and deliver as much information as possible to a single agent scouting the ground. As the agent receives the summaries, it extracts the information about node plates and positions, and tries to reconstruct node trajectories within the area. This is possible by aggregating data related to the same license plate, reported from different summaries.

To determine the effectiveness of MobEyes we decided to evaluate the average uncovered interval and maximum uncovered interval for each node in the field. Given a set of summary chunks related to the same vehicle and ordered on a time basis, these parameters measure respectively the average period for which the agent does not have any record for that vehicle and the longest period. The latter typically represents situations in which a node moves in a zone where vehicle density is low; thus, it cannot be traced by any other participant.

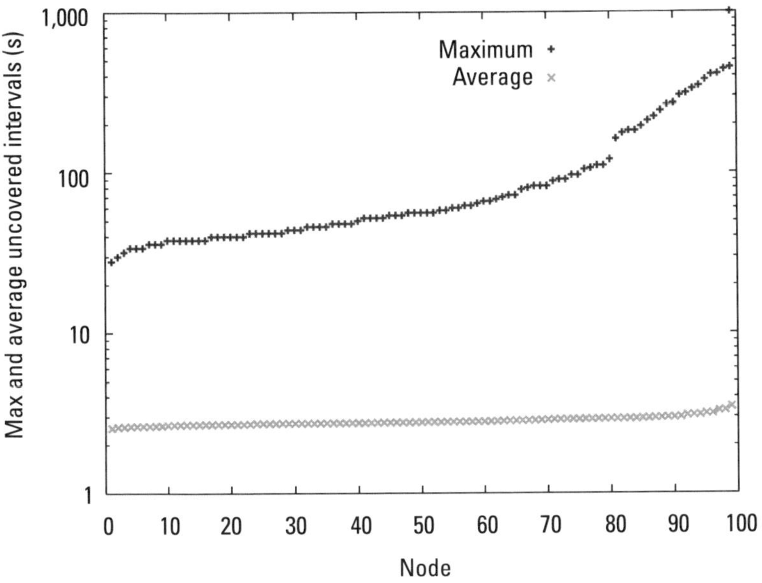

Figure 6.9 Maximum uncovered intervals per node.

We compute the average and maximum uncovered intervals for each simulated node, and present the results in Figure 6.9 (note the logarithmic scale on the *y*-axis). Every point in the figure represents the value of the uncovered interval for the corresponding node. We sorted nodes on the *x*-axis so that they are reported with increasing values of uncovered interval. Results are collected during a 6,000-second simulation. The plot shows that in most cases the average uncovered interval fluctuates between 2.7 and 3.5 seconds; the maximum uncovered interval shows that even in the worst cases the agent has at least one sample every 200 seconds for more than 90% of the participants. A more intuitive visualization of the inaccuracy is given in Figure 6.10. This figure shows, for the case of a node with a maximum uncovered interval equal to 200 seconds (i.e., a node that resides in the worst tenth percentile), the actual trajectory (the unbroken line), and the sample points collected by the agent. As expected, path coverage is worse at the periphery, where vehicular density tends to be lower.

MobEyes can be configured to achieve the most suitable trade-off between latency, accuracy, and overhead by properly adjusting the system parameters, for instance, the k-hop relay scope and the number of harvesting agents. This observation is the starting point for further research activities, like determining the best trajectory of multiple harvesting agents, investigating the optimal interval for summary advertisement, and exploring hybrid strategies that combine broadcast with epidemic dissemination.

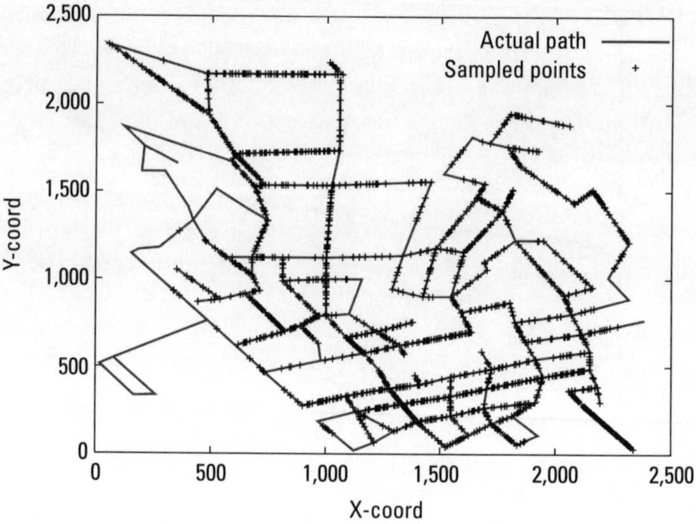

Figure 6.10 Actual node trajectory versus harvested sampled points for the least accurate path.

6.4 Secure, Selective Surveillance in Vehicular Networks

6.4.1 Vehicular Security

Vehicular networks can be used to deploy several different categories of applications as we discussed in Section 6.2. However, the applications that so far have attracted the most attentions are safety related applications. Safety is guaranteed by the periodic exchange of short messages (beacons) between cars that inform them of their relative positions, speed, status of the driver in the car, and possible environmental dangers [42]. There is also safety messaging between cars and infrastructure (e.g., DSRC access points) to report on road perils (say, ice on the bridge, boulder on the road, and so forth) and likewise, there are safety alert broadcasts from access points to cars. Most of the vehicular security work in the past has concerned the security of the safety message exchange. To be more specific, the attacks to vehicles involved in the safe navigation applications can be subdivided in many categories. Hubaux teaches us that attacks can be from insiders (who have been authenticated and possess all the certification material) or outsiders (that have no authentication and thus limited to lower lever, DDoS attacks). The attackers, moreover can be rational (trying to gain personal profit from the attack) or malicious (simply disrupting operations). The attack can be active (generation of messages that confuse or misguide the drivers) or passive (i.e., eavesdropping). A typical attack to vehicles engaged in safety message exchange is the bogus information attack [i.e., active message attack that reports wrong information, for instance to reroute traffic and clear the road in front for the attacker (rational attack) or simply to create chaos among drivers (malicious attacks)]. This attack is an insider attack, because each receiver authenticates the sender before accepting the message. Another well known attack is the Sybil attack, carried out by an inside attacker who pretends to be a multiplicity of legitimate vehicles at the same time, and thus can issue misleading messages to drivers (who in turn try to protect themselves from bogus attacks using a majority consensus). The most common (and feared) passive attack is the location privacy (or location tracking) attack. The attacker, hiding behind access points, say, passively monitors the IDs of passing vehicles and reconstructs their routes. Detailed knowledge of daily driver routines can be used to do serious damage to drivers. To protect from these attacks, the design of the vehicular network must comply with these basic requirements: authentication of the sender; verification of data consistency (by majority rule, say); nonrepudiation (of messages sent); revocation (of credentials); and privacy guarantee. The most popular implementation of these requirements is via public key infrastructure (PKI) [42]. There is, however, a stiff processing and line overhead cost associated with PKI, thus the real challenge of vehicular security (to protect the safety messaging application) is the tradeoff between degree of protection and real-time constraints (I must receive the safety beacon within 100 ms, secure or not; otherwise, I crash).

6.4.2 New Security Requirements for Urban Surveillance

These basic security requirements mentioned above address the safety message application. Additional, more specialized security and privacy requirements arise for nonsafety applications. Let us consider the case of urban sensing and dissemination in a homeland defense application. We must control the massive collection and harvesting of sensitive information from the population of vehicles. For example, a police agent has detected a dangerous radiation leak. He must inform the people around the radiation area and help them evacuate. He does not send the message to drivers far away from the accident, since this may cause large-scale panic, or, it may attract curiosity seekers. On the other hand, he wants to alert all the paramedics and firefighters in a larger surrounding area. Under this situation, it is clear that selective dissemination is needed. Everybody may receive the alert in a large surrounding area; however, only those with specific qualifications (say, firefighters) and/or in a specific area are able to decrypt the message.

The problem at hand is secure group key distribution. It is well known that PKI does not scale well as it would require just too much time and overhead to construct the group from all of its components. This problem is more efficiently solved by an emerging security system called dynamic attribute–based encryption (DABE). It is a security system built on attribute-based encryption and designed for selective and secure group broadcast. In the DABE system, every object is described by a set of attributes. Messages are encrypted by defining the attributes of the authorized receivers without the knowledge of the specific receivers' identities. In the above example, an important attribute will be position (near or far the radiation leak).

A typical use of DABE in a homeland defense (HD) application is the following:

1. *Before the attack:* The HD central post instructs (via WiFi or 3G) trusted vehicles in a certain area of the city to look out for particular suspects. So, they must filter out their camera inputs and send to the HD post only the relevant discoveries. Use DABE to distribute the commands, along with secret key to encrypt response.
2. *After the attack:* The HD central post instructs trusted vehicles in an area to report about a particular accident. It also gives them the credentials to request critical info from regular cars. In other words, HD deputizes trusted cars to collect the critical information. This is more secure and covert than sending a request to all the cars. Again we use DABE to distribute instructions to the trusted cars

6.4.3 Attribute-Based Encryption: A Primer

Attribute-based encryption (ABE) [43, 44] was introduced by Sahai and Waters et al. as a new mechanism for encrypted access control. There are several versions of ABE; the one used to selective broadcast in VANet is so-called cipher-policy attribute-based encryption (CP-ABE) [45].

To some extent, CP-ABE is an extension of conventional PKI to groups: an authority generates public and private keys; the public key is for encryption while users keep their own private keys to decrypt. In CP-ABE, public and private keys are not a one-to-one pair; instead, there is only one public key and a potentially very large number of private keys, one for each user in the target group.

A user's private key is associated with an arbitrary number of attributes. A user acquires a private key from the authority based on his properties. One attribute corresponds to one property. Properties such as name, age, and employer are different from person to person, so users have different attribute values associated with their private keys. A user's attributes associated with his private keys are usually called that user's attributes. Underlying properties of a user may change (e.g., age grows once a year), so do the corresponding attributes. Some attributes change frequently but some do not. Generally attributes that never change is called static attributes; in contrast, dynamic attributes refer to those changing attributes. A user should keep his attributes up-to-date, meaning a user's attributes reflect the user's current state or properties and he is not expected to hold an attribute for which he is no longer (or not yet) certified.

The sender uses a policy tree (i.e., the logical combination of various attributes) and the public key to encrypt a message. Only clients whose attributes satisfy the policy tree can decrypt the message.

A good example to understand how CP-ABE works is a discount coupon, which can be only redeemed if you satisfy a California resident, a University of California or California State University student, a member of AAA or UHaul, and so forth. In the CP-ABE context, the coupon is the information that we must protect, and the restricting conditions are so-called attributes. The message (the coupon) is encrypted with an access control policy tree that contains the logical combination of the different attributes: CA resident AND (UC student OR CSU student) AND (AAA membership OR UHaul member). Any qualified user can apply and obtain a private key from certifying authority. The key is associated with the various qualifications (i.e., attributes) of the applicant. The user can decrypt only if his attributes satisfy the policy tree. As Figure 6.11 shows, two users try to decrypt the message. Kevin has attributes CA resident,

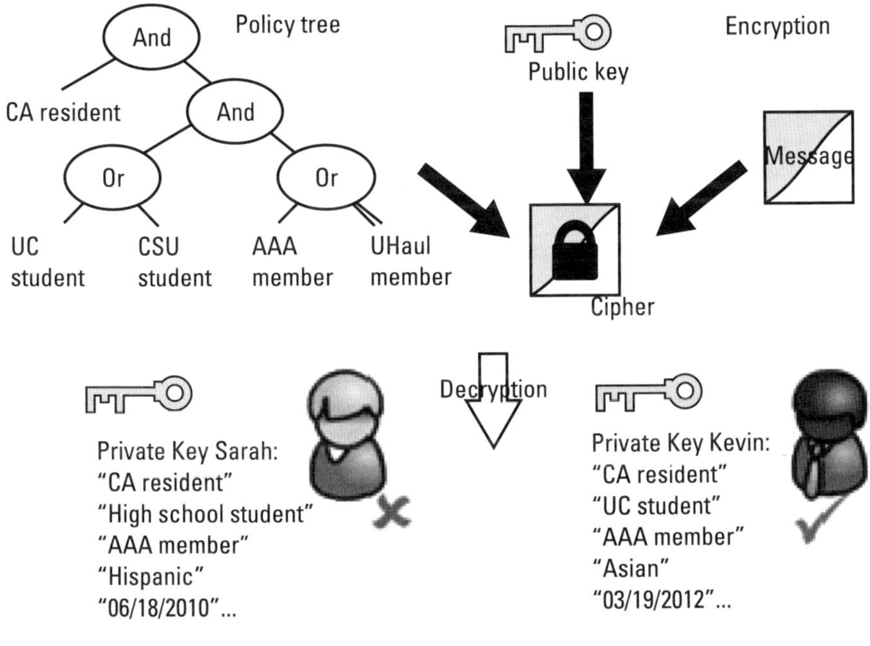

Figure 6.11 An example of a policy tree and private keys associated with attributes.

UC student, and AAA member so that he can decrypt the cipher while Sarah cannot.

6.4.4 DABE: The Better Choice in Dynamic Environments

As cipher-policy attribute-based encryption was initially designed for the general purpose of access control, it did not take much consideration on the real-time constraint, which is one of the most important factors in the VANet and its derived applications. Built on CP-ABE, DABE inherits most of security features and is specifically optimized for dynamic environment, where part of, or all of the attributes are dynamic.

Figure 6.12 is an experiment result of the comparison between CP-ABE and DABE in the dynamic environment. All the attributes are dynamic in the experiment; some change fast and some are relative slower. It shows that when the attributes associated with a private key grow, DABE saves a lot more attribute updates, which is equal to say, DABE system uses less bandwidth and CPU resources than the vanilla ABE system in the dynamic environment.

There are three major components in the DABE system: the key master, senders, and receivers.

As with PKI, the key master is the authority of the security system and takes charge of generating keys, including public and private keys, as well as the attributes associated with private keys. The key master generates private keys and attributes for users according to their requests. It verifies each request

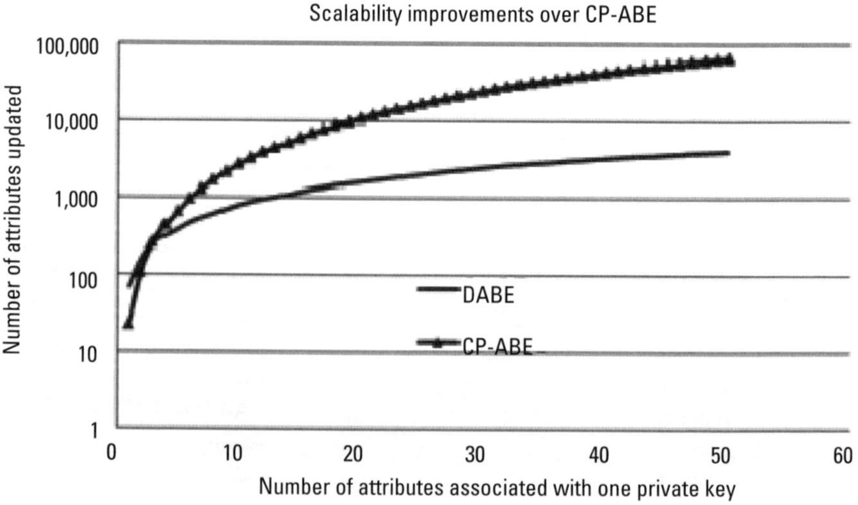

Figure 6.12 DABE versus CP-ABE in a dynamic environment.

for compliance before issuing the key to the user. In addition, the key master is responsible for all the work related to "attributes."

A sender is an individual that encrypts a message and sends it to others.

A receiver receives ciphers and tries to decrypt. A receiver can get a private key before or after receiving ciphers. Generally, to enable "proactive trust," a receiver is supposed to obtain a private key when he enters the system for the first time and keep his private key up-to-date proactively.

All the parties in the system, including the key master, senders, and receivers, are assuming to have coarsely synchronized time clocks using, for example, GPS or equivalent techniques.

1. *Setup:* When the system is initialized, the key master generates and publishes the public key. It also formalizes and maintains the set of attributes that can be used in the system (both in policy trees and private keys), and publishes all the available attributes, their values, and the corresponding descriptions. Each attribute has its lifetime, which is a designed for real-time constraint. When an attribute's lifetime expires, it has to be renewed so that no one can keep an attribute forever. To achieve this, a fading function is introduced to produce a time changing attribute. It can be of any form, as long as it outputs different values for different time inputs, for example,

$$Att\_Current\_Value = Att\_Base\_Value \times \left\lfloor \frac{t - Base\_Time}{60\,\text{min}} \right\rfloor \tag{6.1}$$

   This equation is a fading function with its lifetime equal to 60 minutes. The key master assigns a fading function to each attribute and publishes the fading functions together with the corresponding attribute descriptions and attribute base values.

2. *Encryption:* A sender should know to whom this message is sent when he has a confidential message to be sent out. He abstracts target receivers to a set of attributes and looks up the desired attributes in the key master's attribute list. Then he composes an access policy tree with the attributes downloaded to define the target receivers. After that, he encrypts the message with the policy tree and public key, and broadcasts the cipher.

   There is a little work before the sender can use the attributes downloaded from the key master. As mentioned earlier, a complete attribute published by the key master contains a base value, its description and corresponding fading function. From fading function the sender computes the "latest attribute values," taking attribute base values and current time as inputs. The sender then uses the latest attribute values to compose a policy tree.

3. *Private key generation:* The key master creates a record (a profile) in its memory for each new coming receiver in the system. Records are distinct from receiver to receiver. In addition, the key master keeps this information secret and no receiver will ever know his or others' record.s The key master utilizes records to generate private keys and attributes for receivers.
4. *Decryption:* A receiver requests a private key from key master when he enters the system for the first time, and keeps his attributes up-to-date. As the same in CP-ABE, only if the receiver's attributes satisfy the policy tree with which a message is encrypted, the receiver can decrypt the message. It should be emphasized that attribute A matches attribute B if and only if they have the same attribute value.
5. *Attribute update:* One of the biggest advantages of DABE is that whenever there is a change on the attribute, only that changed attribute is updated while the rest of them associated with the private key are left untouched. When the key master receives the request, it computes the latest attribute value using the appropriate time fading function. Then it looks up the receiver's record and uses the record to generate the requested attribute based on the latest attribute value, and issues the newly generated attribute to the receiver. The receiver adds the new attribute to his private key.

The update is not finished until the old attribute expires. Expiration happens when the attribute value that a sender encrypts with is different from the value based on which a receiver's old attribute is generated. Figure 6.13 explains this process. The sender encrypts a message with two attributes at time t_2. The receiver has both attributes. He obtains Attribute 1 at time t_1 and Attribute 2 at time t_3 respectively. Between t_2 and t_3, Attribute 2's value stays unchanged, which means the key master uses the same value as the sender does, and the receiver can match the sender on Attribute 2. However, from t_1 to t_2, Attribute 1's value has changed. Due to some reason, the receiver has failed to update Attribute 1, meaning this attribute has been expired. So the receiver's Attribute 1 value does not match the sender's value, resulting in his failure to decrypt the message.

6.4.5 Benefits of Dynamic Attribute-Based Encryption

Inheriting the selective broadcast feature from vanilla ABE, DABE has enhanced performance in the dynamic environment: by allowing attribute up-

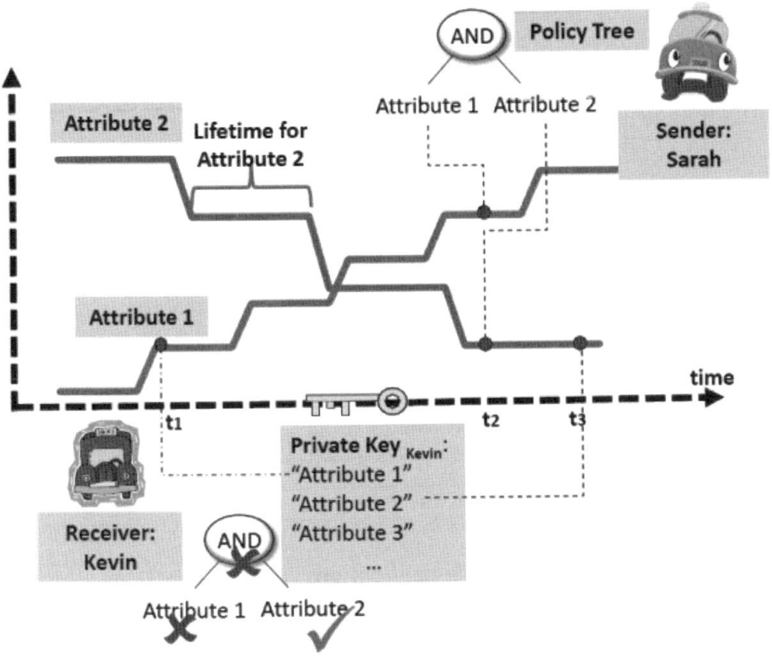

Figure 6.13 DABE prevents a receiver keeping an attribute forever.

dates individually, DABE avoids CP-ABE much more cumbersome private key update (if any of a user's attributes changes, the entire private key, including all the attributes associated with the private key has to be updated), which is demonstrated in Figure 6.12.

Whenever a user's attribute changes, he requests an attribute update. Built on the user's profile, the new attribute is only for that receiver. Due to the uniqueness of each record, receivers cannot exchange their attributes with each other thus preventing collusion, known as the major threat in the ABE security system. On the other hand, this "magic" profile also guarantees that the attribute has no problem associating with that user's private key because all the attributes associated with his private key are generated based on the same record.

The lifetime of an attribute controls the fading speed of that attribute, which is how fast an attribute becomes invalid. If the lifetime is too small, an attribute in the private key set expires very soon after it is issued so that the receiver needs to update it frequently. Too large of a lifetime is not desirable either since "fading" is very important for security consideration: a receiver can use an unexpired attribute to decrypt even after he loses the corresponding property. Generally, an attribute that tends to change frequently, for example, local location attributes (street, neighborhood), should have relatively smaller lifetime to make the old attributes expired as soon as possible. In contrast, an attribute supposed to change less frequently or not change at all, such as high level locations (state, country), may have a bigger value to avoid unnecessary updates. The flexibility of lifetime allows administrators to tune the balance between security and efficiency and build a system optimized for a specific application.

With DABE, trust can be established proactively. If the receiver knows its time/space trajectory, it can request a set of attributes in advance, one for each segment (lifetime). The key master will issue the keys only upon verification that the request is valid (for example, a bus going on a predefined route). On the other hand, DABE is also able to verify the user's past trajectory. When a user moved along a specific route, to keep his attributes up-to-date, he must have collect all the attributes along the route, say Attribute A (for Westwood) at lifetime 2–3 p.m., Attribute B (for Hollywood) at lifetime 3–4 p.m., and Attribute C (for Pasadena) at lifetime 4–5 p.m. So if a receiver is able to decrypt the message encrypted with these three attributes, he is proved to have moved from Westwood to Pasadena from 2–5 p.m.

6.4.6 DABE in Homeland Defense Applications

DABE could have very promising future in the homeland defense security because of its selectiveness and dynamic feature. In a comprehensive homeland defense security system, locations may be encoded into attributes. Roadside units connecting with central certificate authority are responsible for verifying vehicles' location and issue location attributes when vehicles enter a new area. In addition, vehicles have extra attributes describing their properties, especially their belongings, say police agent, firefighter, private taxi, and so forth. The location attributes tend to have a smaller lifetime since vehicles move around quite often while other attributes may have a longer lifetime or even be static, depending on the characteristics of the attributes. All the attributes are updated via the DABE scheme; that is, when a vehicle enters a new location, it requests a new location attribute to keep his attribute up-to-date and the old location attribute expires after a period of time.

Suppose the HD central office is looking out for a particular suspect in the city. The central office knows the suspect will show up somewhere in an area but has no idea about the exact location. It wants all the trusted vehicles to participate in and look for the suspect. In that case, the central office could instruct (via WiFi or 3G) trusted vehicles by sending an alert message encrypted with particular location attributes and attributes defining the set of trusted vehicles. In particular, the attribute tree in the packet header would look like: (policy agent AND firefighters AND private detective) AND (street A) AND (street B) AND (place C). This message contains not only the instructions, such as the appearance and possible behaviors the suspect may have, but also the secret key used to encrypt the response as well. All the vehicles, maybe including the suspect, may receive the message but only trusted vehicles can decrypt. After receiving the instruction, the trusted vehicles must filter out their camera inputs and search for the suspect. If something is found, they may use the secret key to encrypt the response and send it back to the HD central office. HD central office may instruct another message to police agent only near the suspect to take action if they found the suspect, say (policy agent AND street A).

Information collection after an attack is likewise the same as searching for a suspect. The HD central office instructs trusted vehicles in an area to report on a particular accident. The credentials to request critical info from regular cars are attached in the message. In other words, the trusted vehicles are deputized to collect the critical information from regular vehicles. This is more secure and covert than sending a request to ALL the vehicles. In addition, trajectory verification of DABE may be applied to filter reporters who pretend to see the accident or the perpetrators who lies that he is not on spot when accident happens.

6.5 Conclusions

Video cameras are proliferating at a very fast pace in urban areas all over the world, becoming the number one ally of police agent and homeland defense investigators. In this chapter we have shown that fixed video cameras installed in the infrastructure (walls, light poles, bridges, and so forth) have their limitations. We have thus made the case for complementing them with mobile video cameras mounted on vehicles. We have shown that the images and video clips collected by vehicles can be very effectively harvested by law enforcement agents without compromising the privacy of the citizens and without causing undue overhead on on-board processors or on the urban wireless network. We have also uncovered new requirements for secure group distribution of surveillance instructions and for data gathering. These unique requirements have been efficiently addressed using a recently developed cryptographic method called attribute-based encryption. Finally, we have described a campus vehicular testbed that will be used for the implementation, testing, and validation of the many concepts presented in this chapter. We must emphasize that the C-VeT testbed is developed under the NSF Global Environment for Network Innovation (GENI) program. Thus, it represents an experimental environment for testing new ideas and new applications, including mobile surveillance applications. C-VeT will be available for access and experimentation by researchers all over the world.

Today, we can state with confidence that extensive video surveillance supported by vehicles is feasible. The technology exists to properly protect the drivers in terms of security and privacy. We also are aware that a wealth of additional considerations will emerge when such a pervasive system is deployed. For instance, what are the social and psychological implications of constantly observing and being observed? What are the connections with the new phenomenon of Internet social networks, through which we expose the most private events in our life to presumed friends? What is happening to the notion of privacy as we knew it only a few decades ago? These questions are unleashing a wealth of research by social scientists today. We think the results reported in this chapter will help provide an experimental "ground truth" to these new research trends.

References

[1] CameraWatch, http://www.camerawatch.org.uk/.

[2] Stecklow, S., J. Singer, and A. Patrick, "Watch on the Thames," *The Wall Street Journal*, July 8, 2005.

[3] Bulkeley, W., "Chicago's Camera Network Is Everywhere," *The Wall Street Journal*, November 17, 2009.

[4] New York Civil Liberties Union, *Report: Who's Watching?* A Special Report, 2006.

[5] "Can Surveillance Cameras Trap Terrorists?" *The Week*, May 4, 2010, http://theweek.com/article/index/202577/Can_surveillance_cameras_trap_terrorists.

[6] Babwin, D., "Chicago's Ubiquitous Surveillance Cameras Have One Catch: Criminals," *Dallasnews*, April 7, 2010.

[7] Karkils, L., "MAP: Surveillance in the City," *The Washington Post*, February 11, 2008.

[8] "ISIS: New Video Camera Sees It All," http://www.dhs.gov/files/programs/gc_1273160563362.shtm.

[9] Lee, U., et al., "Efficient Data Harvesting in Mobile Sensor Platforms," *IEEE PerSeNS*, Pisa, Italy, March 2006.

[10] MIT's CarTel Central, http://cartel.csail.mit.edu/.

[11] Hull, B., et al., "CarTel: A Distributed Mobile Sensor Computing System," *ACM SenSys*, Boulder, CO, October–November 2006.

[12] InternetCAR Project. http://www.icar.wide.ad.jp.

[13] Ernst, T., K. Mitsuya, and K. Uehara, "Network Mobility from the InternetCAR Perspective," *Journal of Interconnection Networks*, Vol. 4, No. 3, September 2003.

[14] FleetNet, http://www.et2.tu-harburg.de.

[15] Enkelmann, W., "FleetNet—Applications for Inter-Vehicle Communication," *IV'03*, Columbus, OH, June 2003.

[16] Eisenman, S. B., et al., "MetroSense Project: People-Centric Sensing at Scale," *ACM WSW*, Boulder, CO, October–November 2006.

[17] Riva, O., and C. Borcea, "The Urbanet Revolution: Sensor Power to the People!" *IEEE Pervasive Computing*, Vol. 6, No. 2, 2007.

[18] Dikaiakos, M. D., et al., "VITP: An Information Transfer Protocol for Vehicular Computing," *ACM VANET*, Cologne, Germany, September 2005.

[19] Burke, J., et al., "Participatory Sensing," *ACM WSW*, Boulder, CO, October–November 2006.

[20] Juang, P., et al., "Energy-Efficient Computing for Wildlife Tracking: Design Tradeoffs and Early Experiences with ZebraNet," *ACM ASPLOS-X*, San Jose, CA, October 2002.

[21] Small, T., and Z. J. Haas, "The SharedWireless Infostation Model—A New Ad Hoc Networking Paradigm (or Where There Is a Whale, There Is a Way)," *ACM MOBIHOC*, Annapolis, MD, June 2003.

[22] University of Dartmouth MetroSense, http://metrosense.cs.dartmouth.edu/.

[23] Shah, R. C., et al., "Data MULEs: Modeling a Three-Tier Architecture for Sparse Sensor Networks," *Elsevier Ad Hoc Networks Journal*, Vol. 1, No. 2–3, September 2003, pp. 215–233.

[24] Li, Q., and D. Rus, "Sending Messages to Mobile Users in Disconnected Ad-Hoc Wireless Networks," *ACM MOBICOM*, Boston, MA, August 2000.

[25] Wang, Y., and H. Wu, "DFT-MSN: The Delay/Fault-Tolerant Mobile Sensor Network for Pervasive Information Gathering," *INFOCOM'06*, Barcelona, Spain, April 2006.

[26] Gibbons, P. B., et al., "IrisNet: An Architecture for a Worldwide Sensor Web," *IEEE Pervasive Computing*, Vol. 2, No. 4, October-December 2003, pp. 22–33.

[27] Nath, S., J. Liu, and F. Zhao, "Challenges in Building a Portal for Sensors World-Wide," *ACM WSW*, Boulder, CO, October–November 2006.

[28] CENS' Urban Sensing, http://research.cens.ucla.edu/projects/2006/Systems/Urban Sensing/.

[29] Dlagnekov, L., and S. Belongie, Recognizing Cars, Technical Report CS2005-0833, *UCSD CSE*, 2005.

[30] Bellavista, P., et al., "Standard Integration of Sensing and Opportunistic Diffusion for Urban Monitoring in Vehicular Sensor Networks: The MobEyes Architecture," *IEEE ISIE*, Vigo, Spain, June 2007.

[31] ns-2 (The Network Simulator), http://www.isi.edu/nsnam/ns/.

[32] Zhou, B., K. Xu, and M. Gerla, "Group and Swarm Mobility Models for Ad Hoc Network Scenarios Using Virtual Tracks," *IEEE MILCOM*, Monterey, CA, October–November 2004.

[33] U.S. Census Bureau, "TIGER, TIGER/Line and TIGER-Related Products," http://www.census.gov/geo/www/tiger/.

[34] Bai, F., and A. Helmy, "Impact of Mobility on Mobility-Assisted Information Diffusion (MAID) Protocols," *Technical Report*, USC, July 2005.

[35] Broch, J., et al., "A Performance Comparison of Multi-Hop Wireless Ad Hoc Network Routing Protocols," *ACM MOBICOM*, Dallas, TX, October 1998.

[36] Saha, A. K., and D. B. Johnson, "Modeling Mobility for Vehicular Ad Hoc Networks," *ACM VANET'04*, Philadelphia, PA, October 2004.

[37] Rappaport, T., *Wireless Communications: Principles and Practice*, New York: IEEE Press, 1996.

[38] Zhao, J., and G. Cao, "VADD: Vehicle-Assisted Data Delivery in Vehicular Ad Hoc Networks," *IEEE INFOCOM*, Barcelona, Spain, April 2006.

[39] Da Chen, Z., H. T. Kung, and D. Vlah, "Ad Hoc Relay Wireless Networks over Moving Vehicles on Highways," *ACM MOBIHOC*, Long Beach, CA, October 2001.

[40] Bettstetter, C., G. Resta, and P. Santi, "The Node Distribution of the Random Waypoint Mobility Model for Wireless Ad Hoc Networks," *IEEE Transactions on Mobile Computing*, Vol. 2, No. 3, July-September 2003, pp. 257–269.

[41] Lee, U., et al., "FleaNet: A Virtual Market Place on Vehicular Networks," *V2VCOM'06*, San Jose, CA, July 2006.

[42] Maxim Raya and Jean-Pierre Hubaus, "The Security of Vehicular Ad Hoc Networks," 3rd ACM Workshop on Security of Ad Hoc and Sensor Networks (SASN), 2005, Alexandria, VA, 2005.

[43] Sahai, A., and B. Waters, "Fuzzy Identity Based Encryption," *Advances in Cryptology C Eurocrypt*, Vol. 3494 of LNCS, Springer, 2005, pp. 457–473.

[44] Goyal, V., et al., "Attribute Based Encryption for Fine-Grained Access Control of Encrypted Data," *ACM Conference on Computer and Communications Security (ACM CCS)*, 2006.

[45] Bethencourt, J., A. Sahai, and B. Waters, "Ciphertext-Policy Attribute Based Encryption," *Proceedings of the 28th IEEE Symposium on Security and Privacy*, Oakland, CA, 2007.

[46] Peterson, L., et al., "A Blueprint for Introducing Disruptive Technology into the Internet," *Proceedings of HotNets–I*, Princeton, NJ, October 2002.

[47] Kotz, D., and T. Henderson, "Crawdad: A Community Resource for Archiving Wireless Data at Dartmouth," *IEEE Pervasive Computing*, Vol. 4, No. 4, 2005, pp. 12–14.

[48] Clausen, T., and P. Jacquet, "Optimized Link State Routing Protocol (OLSR)," *RFC3626*, 2003.

Appendix 6A The C-VeT Testbed

It is important to verify the feasibility of pervasive vehicular surveillance and its efficient coexistence with infrastructure video cameras in an experimental setting. In this Appendix we describe C-VeT, the vehicular testbed currently deployed at UCLA to assist us in evaluating several advanced vehicle applications including home defense video surveillance.

6A.1 The C-VeT Architecture

Our vision is for C-VeT to be an open platform to support vehicular network and urban sensing research and related applications. We were inspired by the pioneer work done by Larry Peterson and Tom Anderson with Planet Lab [46]. In particular we have built an always on, fully virtualized, Web-accessible, sensor equipped testbed infrastructure. The UCLA campus, with its 10 acres of urban development, reproduces many of the scenarios, propagation, and communication challenges typical of a city, in a realistic manner but yet relatively small-scale. In particular, the C-VeT architecture provides:

- A fully virtualized platform that runs both Linux-based and Windows-based operating systems with full insulation among the guest virtual machines, and enables the users to redesign low-level protocols such as, for instance, MAC protocols. This feature is key for network centric experiments.
- A campus-wide mesh network, developed using OPEN WRT and optimized for the integration and support of the vehicular network. It helps to cope with network disruptions (quite common in small-scale testbeds), and enable opportunistic, interactive, and delay tolerant experiments that exploit the infrastructure.
- Thirty facility management vehicles equipped with the C-VeT hardware/software, providing an always-on platform to run experiments, and collect traces and measurements. The facility management vehicles perform both routine maintenance trips and on-demand interventions, in response to emergencies resulting in a varied mobility pattern that well approximates real city traffic.
- Thirty commuting vans, equipped with the C-VeT-Census platform that survey the environment gathering traffic and air quality, and stereoscopic images. The aim is to build a large micropollution database that enable new models and facilitates also visual environment surveys (see Figure 6A.1).

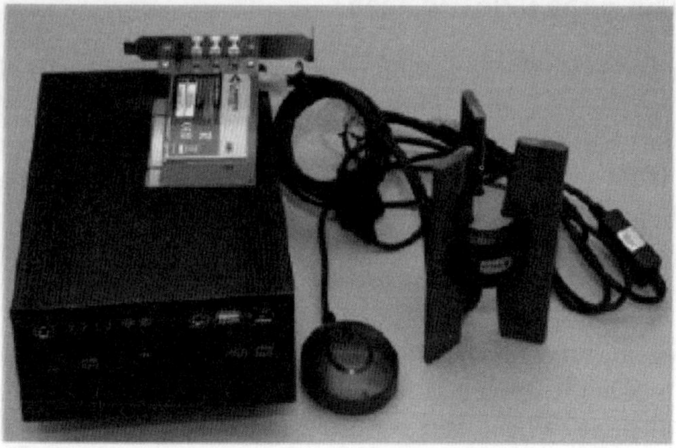

Figure 6A.1 C-Vet mobile node.

- A number of downloadable preconfigured virtual appliances, to allow users to develop the protocols to be tested at home with a compatible software configuration.
- A large-scale emulator that allows users to debug their algorithms and protocols on the same hardware as the actual C-VeT nodes, but with an emulated network component developed with the Qualnet hybrid simulation.
- A robust Web interface, able to manage the users and deploy the experiments in a streamlined fashion. The Web server provides the front-end for a number of user friendly services and tools, enabling users to focus on research rather than testbed implementation. For example: services to set up the experiments and gather the data; APIs to low lever interfaces for hardware component virtualization; and virtual MadWiFi layer for the support of virtual machines.
- An organized live database of mobility traces; sensed environmental data; road traffic information; vehicle CanBus statistics; MAC layer statistics (through MAD WiFi); and physical layer statistics, taken using a variety of radios (cognitive radios, MIMO, and so forth). This data collection is made available to the research community in collaboration with existing trace collection programs and archives, such as CRAWDAD [47].

We have designed the testbed using a top-down approach and the whole system can be described through a number of relatively simple building blocks: the C-VeT mobile node, C-VeT mesh node, C-VeT-Census platform, Web-based control center, and emulation platform.

The C-VeT infrastructure is designed to provide an always-on facility for research in a wireless vehicular network. To achieve this goal we chose to install our equipment in the UCLA campus facility management and van pool vehicles. Those cars and vans are driven everyday to fulfill the campus needs and perform both routine and nonroutine tasks. In addition to the permanent facility vehicles, there is a small pool of private vehicles equipped with C-VeT nodes that can be driven by the researchers themselves for customized, controlled experiments.

The C-VeT mobile node (Figure 6A.1) is an industrial strength Cappucino PC powered by an Intel Dual Core Duo processor at 2.5 GHz, 2 GB of RAM, 320 GB disk. Hard drive and internal parts are rugged to sustain physical stress (i.e., large temperature fluctuations, vibrations, and so forth). The PC has three wireless interfaces: IEEE802.11a/b/g/n based on the Atheros AR9160 chipset; IEEE802.11p interface based on a Daimler-Benz customized chipset; a standard Bluetooth interface mostly for internal communications. Other radios can also be retrofitted in the mobile node platform. In particular, a few vehicles may be equipped with programmable Silvus SC2000 MIMO platforms (4×4 configuration) that provide full access to the physical layer and enable a new generation of experimental MAC layer research.

The C-VeT nodes are instrumented with a customized sensor platform designed to provide a flexible data collection. This includes infrared-based CO_2 sensors, electrochemical CO sensors, SIRF III or Ublox-based GPS sensors, temperature and humidity sensors, and a megapixel camera. Using the C-VeT cars as mobile air quality sensors enables a new wave of atmospheric research, aimed at the use of mobile sensing agents, to study the air quality at the neighborhood. Part of the fleet features high-performance exhaust particulate sensors DC2000CE, thus being the first testbed able to support the currently leading research in microclimate air quality.

The C-VeT mesh node is based on MobiMESH hardware. The C-VeT mesh nodes feature Open WRT OS and Atheros Chipset with MadWiFi support, thus easing up the integration with mobile nodes. The fixed infrastructure is planned to be installed on the rooftops of UCLA buildings, aiming at full campus coverage and integration with the existing campus WiFi infrastructure. The mesh allows opportunistic Internet access from the vehicles, and provides also a control channel to the vehicles. The mesh network can be configured via the Web (e.g., customized routes can be set up by the network operator to perform particular experiments). This C-VeT integrated approach with infrastructure and vehicles broadens the experimental scenarios. In the initial phase we will cover the south campus and create an initial backbone of six mesh points. The initial campus coverage map is shown in Figure 6A.2.

6A.2 MobiMESH

MobiMESH networks feature multiradio mesh routers, a provisioning, management and monitoring system, and security procedures, thus providing a complete platform for wireless mesh networking, which can be employed in a wide variety of environments/applications. The following characteristics of MobiMESH hardware/software solutions make them particularly suited for the C-VeT vision:

- *Broadband backhauling:* MobiMESH networks are able to build up and dynamically maintain a broadband wireless backbone that can be used to support/complement vehicle-to-vehicle communications.
- *Mobility support:* Wireless devices are allowed to seamlessly roam within MobiMESH networks without losing active connections.
- *Flexibility:* MobiMESH networks are self-configuring and self-managing.

Since MobiMESH is the core of C-VeT on the backbone's side, we give hereafter a concise technical description of MobiMESH solutions.

MobiMESH features a hybrid mesh network architecture. Indeed, the network consists of three main architectural building blocks, as shown in Figure 6.A3:

Figure 6A.2 C-VeT infrastructure.

- A mesh backbone composed of MobiMESH wireless mesh routers, which provide the routing and mobility management infrastructure, and is further connected to gateways;
- An ad hoc extension, which is responsible of extending MobiMESH functionalities to mobile nodes;
- An access network, which can be used by standard WiFi clients to get connectivity.

The mesh backbone and the ad hoc extension are based on the ad hoc network paradigm, where all nodes and mesh routers collaborate to routing traffic; routing is provided through a proactive ad hoc routing protocol based on OLSR [48], and properly modified to account for multiple radios at the mesh nodes, and link quality metrics. The backbone network is also responsible for the integration with the wired network, through devices called gateways, which are equipped with a wired interface and that can route traffic to the wired network and therefore to the Internet.

The access network is rather flexible and operates in the infrastructure mode, so that standard clients perceive the network as a standard WLAN and

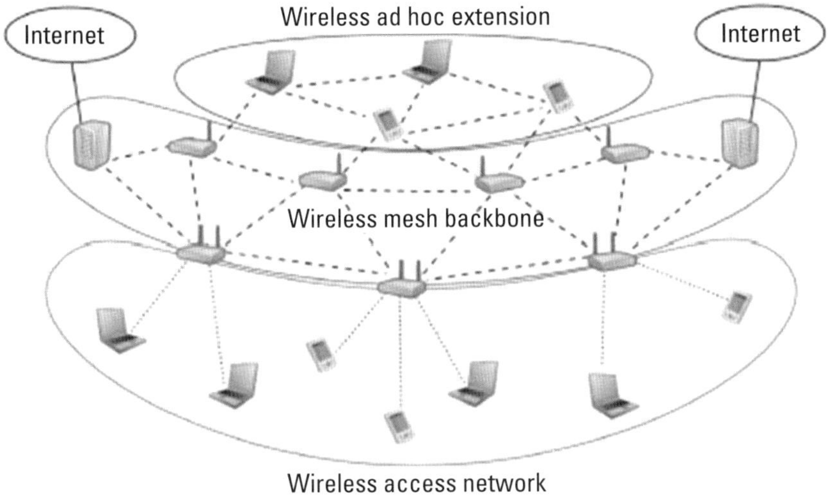

Figure 6A.3 MobiMESH network architecture.

behave accordingly; in this way MobiMESH can also be accessed by standard WLAN clients with no specific software installed.

The MobiMESH mesh routers represent the main building block of the MobiMESH network, since they are responsible for creating the broadband backhaul, further offering access to wireless mobile clients. The MobiMESH mesh routers can be equipped with two to four radio interfaces, which can be flexibly used either as backbone or access interfaces. Moreover, any interface can be tuned to any available channel in the two frequency bands 2.4 GHz, 5.7 GHz, and 5.9 GHz (via DSRC). Mesh routers, which have also an interface dedicated to WiFi access, are called access routers.

7

Detection and Identification of Dangerous Materials for Airport Security

Pierfrancesco Lombardo, Fabiola Colone, and Roberta Cardinali

7.1 The Airport Transport Security Problem

The problem of air transport security has been dramatically presented to the public opinion, as well as to experts and governments, by the events of September 11, 2001, followed by the August 2006 disruption by Scotland Yard of the terrorist plan to use liquid explosives carried on board several aircrafts traveling from the United Kingdom to the United States and Canada.

As apparent to the reader, the airport security is at the same time very important and very difficult to address. Its importance is quite clear because:

- The large amount of people simultaneously located in the airport areas can represent a natural target for terrorist acts and other forms of crime.
- Similarly, the high concentration of people on large airliners, with no escape exit, can be considered a favorable terroristic target.
- The potentially high level of damage to people's lives caused by attacks on aircraft makes them a relatively easy tool to produce high effects with limited effort.

- The impact of the terroristic action on the world economy, as a side effect of the personal fears from potential passengers, can become extremely large.
- The recent trend of using hijacked airplanes as lethal weapons to be used also against civil targets provides an easy and effective means of implementation for terrorist actions.

The perception of the difficulty in providing an effective and reliable airport security can be easily obtained by considering, among the several relevant factors, the wide scope of the protection actions to be provided, and in particular:

- The wide extent together with the highly nonhomogeneous characteristic of the areas to be monitored;
- The wide range of threat types to be faced (liquid or solid explosives, metallic and nonmetallic objects, guns, and so forth);
- The high concentration of people with continuously changing positions and densities;
- The heterogeneous typology of the persons located in different airport areas (screened passengers, airport personnel, and unscreened passing persons).

Although airport security has always been considered a major priority for both the aviation industry and national governments, airport security measures have experienced a number of failures. In particular, many harmful terrorist actions followed the adoption of ineffective protection measures that were unable to stop people from carrying explosive materials, guns, or items usable as weapons inside the airports and/or aboard aircraft.

As a basis for the EU FP7 Project ATOM [1], which is introduced in some greater detail later, an analysis was performed of the RAND–MIPT Terrorism Incident Database [2]. A list of 225 attacks since 1980 on civilian airplanes or airports worldwide, including 150 attacks on civilian airplanes and 75 attacks on or at airports, was obtained. The 150 recorded attacks on aircraft caused 4,280 fatalities since 1980, compared with just 76 fatalities resulting from attacks on airports. While this number goes down to about 1,400 when excluding the casualties of the September 2001 attacks, it is still about 20 times larger than the number for attacks on airports. Although the attacks on airplanes have certainly been found to be much deadlier, the 75 attacks on airports have also produced significant damage:

- Forty-nine used portable explosives, mostly in luggage but also including three mail bombs (58 fatalities);
- Nine used vehicle bombs (four fatalities);
- Eight used missiles, rocket-propelled grenades, or mortars (one fatality);
- Seven used firearms (13 fatalities).

A list of the most notable cases of tragic consequences of the failure of airport security measures is reported in Appendix 7A, taken from the preparatory work for the ATOM Project [1]. An unknown number of criminal, nonterrorist attacks, where firearms or other small weapons were used, should be added to this analysis.

Despite the tighter security measures adopted after September 11, 2001, (with increased passenger wait times), the U.S. Transportation Security Administration (TSA) reported that in 2002 roughly 60% of fake bombs or component parts to bombs were missed by covert screeners during dummy tests carried out in several major airports. In 2007, that percentage rose to 75% [3]. This is confirmed by Appendix 7A, which lists a number of major actions taking place after 2001.

As apparent from these considerations, the air transport security problem includes protection from attacks of both airports and departing airplanes. However, the protection measures for both aspects of the problem need to take place in the airport area:

- Airport protection is meant to provide the security of the many people that every day depart from, arrive, work, or wait there for arriving or departing people.
- Departing airplane protection is meant to avoid people bringing weapons or bombs into the airport, thus largely reducing the probability for these items to reach the airplane.

For these reasons the airport area is the focal point for the air transport security system.

In the following sections we introduce the state of the art of the airport security measures and describe some of the innovative solutions presently under study and development.

7.2 Standard Security Solutions and Issues

The standard airport security solution is obtained by segmenting the airport area into three areas:

- The *public area*, which includes all the ground side areas of the terminal and extends to the screening checkpoints that mark its border. Anyone can enter this area and no assumptions can be made about the security level of people and objects present inside it.
- The *sterile area*, which begins from the screening checkpoints and includes all of the air side of the terminal, as well as the interior spaces of the gates, the jet ways, and the aircraft. All people, carry-on bags, and possible objects present in this area are assumed to have been carefully screened and thus to be free of illicit, dangerous, or other forbidden items.
- The *off-limits area*, corresponding to the airport region outside the passenger lounges, which includes the aprons, taxiways, runways, and the remaining areas in communication with the airplanes' exterior and luggage holds. This area is off-limits for both passengers and visitors; only airport employees with appropriate authorization are admitted in it.

The standard procedure assumes that people move freely in the public area (where everything is officially classified as unsafe) and are subject to an intensive and accurate screening at the checkpoints before being admitted to the sterile area (where everything is assumed to be safe), which includes all the gates to be embarked on a departing flight. When disembarking from an airplane, passengers are directly admitted to the sterile area, so that they do not need to be rescreened if disembarking from a flight arriving from a national or secure departure airport. Additional measures can be used to increase confidence in the results, such as adding extra screening for transfer flight passengers or having specially trained individuals who engage passengers in a conversation to detect threats rather than solely relying on screening equipments to find threats. An even more secure solution is only used for identification and screening of airport and aviation personnel, especially for the ones authorized to enter the off-limits area.

This standard solution is essentially and mainly centered on the checkpoint, which is to any extent the place where most of the security measures are applied, including the screening of carry-on luggage and passengers. Therefore, its effectiveness is essential for global airport security.

At present, hand luggage screening is applied just before passenger screening by means of a conveyor belt that moves it through the X-ray device. This device provides very detailed images that allow the identification of potentially nonauthorized items, along with their position inside the luggage. The hand luggage search by a member of the security staff (which is perceived by passengers as a reduction of privacy) is limited to the specific item position, thus reducing both the search time and the extent of the hand luggage search. Advanced technology X-ray equipment that typically exploits multiple views together with higher resolution [4] is being introduced. Machine operators are trained to look for suspicious items, and not just obviously suspicious items like guns or knives, but also anything that could be a component of an improvised explosive device (IED). Since there is no such thing as a commercially available bomb, IEDs are the way in which most terrorists and hijackers penetrate controls. An IED can be made in an astounding variety of ways, from basic pipe bombs to sophisticated, electronically-controlled component bombs. Many airlines have set up explosive detection and screening (EDS) machines in airport lobbies near check-in stations; others use inspectors or dogs to manually screen bags. New devices have also been introduced to detect traces of explosives using samples taken by contact or by puffs of air, under the name of explosive trace detectors (ETD) [5]. Both methods have their drawbacks. Manual examination, besides requiring time, is a tedious process, and can sometimes introduce error. EDS, whose operational principle is based on the detection of volatile compounds given off from explosives using a kind of gas chromatography, take up space otherwise available for passengers or create a crowded environment. Besides, it should be considered that the bombers often use analog commercial sniffers and gateways aimed to evade trace detection to help them to test their devices.

Additional tests are performed on electronic devices that are switched on, at least to check that they are not just cases used to hide unauthorized items. Special devices are being developed for the analysis of passengers' shoes potentially carrying explosives; one of these special devices is a quadrupole resonance carpet, which detects threats in shoes without having to take them off.

Passenger screening cannot be performed using standard X-ray technology (namely high energy absorption X-ray technology), which requires a type and level of radiation not safe for human beings and is generally applied by means of walk-through metal detectors. Unfortunately, a metal detector does not give any indication of the shape and material (and thus type) of the object causing the alarm, nor does a metal detector give any indication of the position of the detected object on the passenger's body. As a consequence, coats and jackets must be removed before entering the metal detector gate and all objects must be removed from the passenger's pockets; moreover, a global passenger search is required if the metal detection gate goes off. In addition, its sensitivity is not very high and largely depends on the material, so that not all objects are detected.

Recently, X-ray backscatter imaging systems have been introduced and tested in the United States. They still use ionizing X-ray technology, but at very low levels of energy, which is considered by the U.S. Transportation and Security Administration (TSA) to be safe for the human body. These devices measure the radiation reflected by the bodies, thus allowing scanners to obtain detailed full-body images of passengers and identify potentially concealed objects of different types, since various materials have sensibly different characteristics in terms of X-ray beam deflection [5].

A new generation of millimeter-wave devices has been recently introduced; these devices are able to produce images of body contours and objects of different materials present close to it, thus showing similar characteristics to X-rays devices. Although they use the largely safe technology of radiations that are reflected off the skin, they are much more sensitive than metal detectors and are specifically able to identify nonmetallic objects. Moreover, these devices are able to provide a clear indication of the position of the detected item. The technology has been recently tested in Europe. Introduced under the name of Security Scan for a voluntary test before the standard application at Schiphol Amsterdam Airport as a joint initiative of National Counter-Terrorism Coordinator (NCTC), Customs authorities, and Schiphol Amsterdam Airport [6], these devices show a much higher effectiveness than metal detectors. Although providing detailed images of the body underneath clothes raises privacy issues (see for example, [7]), both millimeter-wave and X-backscatter imaging sensors, which are also addressed as body scanners, can be considered to provide a significant increase in the level of privacy with respect to the use of metal detectors, because only an expert image analyst examines the images in a closed space and in case of an alarm the passenger search is limited to the area where the object has been detected. Similarly to X-ray backscatter imaging systems, Security Scan still requires a person to enter a special box, assume a good position for screening, and wait for few seconds of screening time; nonetheless, it speeds up the overall process by reducing the hand searches. While already in use in its first versions, the millimeter-wave imaging sensors will be considered in more detail in Section 7.4, since it is expected to gain wide acceptance with its improved versions and plays a significant role inside the presented innovative solution.

Noticing the demand for new technology in airport security, in late December 2005 GE Security unveiled what it called its Checkpoint of the Future, incorporating an array of futuristic scanning technologies: (1) automated carry-on scanning, (2) automatic biological pathogen detection, (3) millimeter-wave full-body scanning, (4) explosives trace detection on passengers, and (5) quadrupole resonance carpet. In January 2007 three of its scanning technologies were included in a product called the Secure Registered Traveler (SRT) Kiosk, which is now in use at Orlando International Airport, three terminals at JFK Airport, Indianapolis, Cincinnati, Northern Kentucky, San Jose, Albany, New York, and Little Rock, Arkansas, and was about to go live at the time of this writing at Newark, LaGuardia, and a few other airports. These new sophisticated technologies can certainly bring an improvement in airport efficiency; however, the human factors involved with the security personnel training and operating conditions are also significant parts of the picture [8].

In addition, the use of various types of video cameras (fixed, pan-tilt-zoom, and so forth) that have become almost ubiquitous for many applications is largely exploited inside airports for surveillance purposes, inside both the public area and the sterile area. Despite their inability to detect concealed items and identify objects composition, video cameras can be very useful to detect, track, and recognize people and objects. While not restricted to checkpoints or to specific areas, cameras can also be of major importance in the process of identifying anomalies that require special attention. A large effort is being made recently to get the maximum benefit from the joint use of large networks of multiple smart cameras, to move from the projective nature of these imaging sensors to the reconstruction of a fully dynamic 3-D view of wide areas able to provide increased situational awareness for airport security (see, for example, [9]). For a more detailed review of the present and foreseeable situation for passenger and baggage screening, the reader is referred to [10, 11], which present a global view originally prepared for the U.S. Congress.

To evaluate the quality of the airport security process and summarize the issues emerging from these considerations, four main performance measures can be introduced:

1. *Quality of the screening:* The capability of the security system to detect the items of the prohibited list [12], with an acceptable number of false alarms. Typical performance parameters are the detection probabilities for each item in the list for an assigned false alarm rate.
2. *Process perception:* The perception that both passengers and security personnel have of the security process. This includes, in particular:
    - The process provides a high-quality screening (a typical performance parameter is the percentage of perceived reliability of the screening process).
    - Personal privacy is respected (a typical performance parameter is the percentage of privacy respect).
    - The waiting time is acceptable (a qualitative performance parameter is typically used to express the level of acceptability).
3. *Process time:* The time required to perform the screening process.

4. *Costs:* The cost of the security in terms both of equipment and security personnel.

At present, it is very difficult to set absolute figures for the performance parameters that determine the acceptability of a security system. A process is ongoing by TSA in the United States and the regulatory entities in Europe to attempt to define globally acceptable performance requirements. In contrast, these parameters are quite useful to assess the potentialities of different sensors and systems and to compare them to each other. Clearly, large improvements are still being measured moving from the metal detectors, unable to detect nonmetallic objects, to body scanners; similarly, wide improvements are being made in automatic sniffers and for explosives and automatic IED detectors with respect to operator random sampling only.

From these considerations we can conclude that, at present, X-rays make carry-on luggage screening rather effective, since they allow the identification of a large number of objects and their location, so that the carry-on luggage search is reduced to a minimum, thus respecting privacy, reducing process time, and providing the perception of an accurate security control. However, still the need to queue at the checkpoint gives the perception of a non-negligible waiting time.

Passenger screening raises a much larger number of issues about effectiveness, privacy, and waiting times. The introduction of millimeter-wave scanners appears to increase the quality of the screening compared to the standard metal detectors, which were only effective for metallic objects, and the level of privacy perception. In fact, nonmetallic objects are also detected and their contours are clearly visible. Still, the quality of the scan needs to be increased to demonstrate the effectiveness against all items in the list [12]. Imaging at a high resolution also improves passengers' perception of the screening quality. However, a large collaboration of the passengers is still required, and a short but nonnegligible amount of time might be required due both to the imaging time and to the potential hand search. This also applies to the complex multisensor systems that certainly bring an improvement in airport efficiency (for example, passengers won't have to remove shoes or coats during the screening, nor cell phones or laptops from the bags), and also provide a more detailed search that is less intrusive to the passengers, but still requires passengers to pass through special check points.

7.3 Innovative Approach to Security

As apparent from the previous section, the main challenges of the present security system are related to:

- Increasing the screening quality;
- Strongly reducing the waiting time for passengers (and also carry-on luggage) screening;
- Increasing the perception of privacy through the security system.

While the increase of specific technologies (as, for example, millimeter-wave imaging capabilities) is expected to produce a significant benefit for all considered items, a main issue remains the fact that all the screening is performed in a unique and specific point, namely the checkpoint that separates the public area from the sterile area. Although it is effective to provide an area where people can be assumed to be safe, it forces people to queue at the checkpoint to be screened together with their carry-on luggage, which has a strong impact on the flow of people though the airport. In addition, we cannot avoid noticing that, with this approach to security, there is no control of what happens in the public area, which is largely unprotected from potential terrorist attacks. Therefore, it would be reasonable to start the screening process well before the passengers enter the gate area.

Within the framework of project ATOM (Airport detection and Tracking Of dangerous Materials by passive and active sensors arrays) [1], the applicants propose a more efficient security screening based on the nonintrusive detection of threats (explosives or arms) and the tracking of threatening people. The aim is to study, design, and develop the functional prototype of an innovative multisensory-based system integrating active and passive radar sensors, able to survey wide airport areas without requiring the passengers' cooperation by detecting prohibited items and tracking the movements of people carrying these items. The main characteristic of the ATOM system is to extend the area under surveillance outside the gate area, including in the security screening all people in the airport areas (passengers, airport personnel, visitors).

This security system is intended to be nonintrusive but pervasive, on the basis of the integration of active and passive radar technology. The nonintrusiveness feature is due to the fact that in principle the system does not require ad hoc checkpoints through which passengers will have to pass to be screened. Such a system aims at providing not only surveillance of airport terminal area without the collaboration of the passengers, but also an improvement of the security before the gate area entrances.

Since the detection and identification of dangerous items is not done only at very specific points (namely the checkpoints), now it is necessary to provide the operator with the tracking of the person carrying the illicit object, in order to facilitate the counteraction, by allowing the security personnel to reach the terrorist. This function, which is essential in a distributed detection and identification security system, is not required in a checkpoint-centered surveillance, where it is assumed that there will be an immediate stop of the suspicious passenger at metal detector (or security scan). Moreover, it becomes important to the integration of all the surveillance activities in an advanced control and command system in order to be able to react and to improve the situation awareness and the crisis management. A pictorial view of this scenario is depicted in Figure 7.1.

Following this approach, the scope of the innovative security system is threefold:

- Detection and identification of prohibited items concealed under clothes or inside bags, without interfering with the passenger flows and without requiring passengers to remove their clothes or empty out their bags;
- Tracking of the movements of the passengers carrying prohibited items;
- Data management and distribution for handling the various information flows and enabling constant tracking of people carrying prohibited items while providing real-time feedback to security personnel in the security control center or within the airport area.

The approach to be followed foresees two separate and integrated sensors systems (see Figure 7.2).

The first sensor system is mainly located around the terminal accesses and is equipped with innovative active devices, not interfering with passengers'

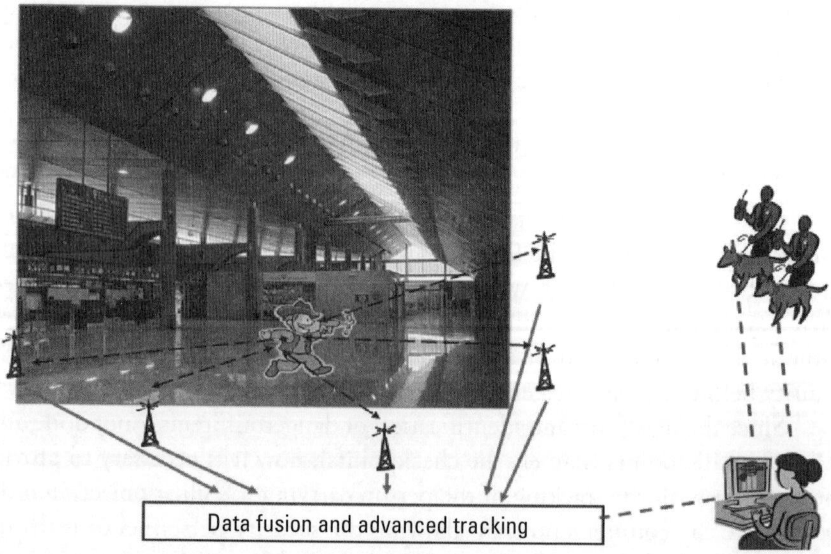

Figure 7.1 The distributed security system scenario.

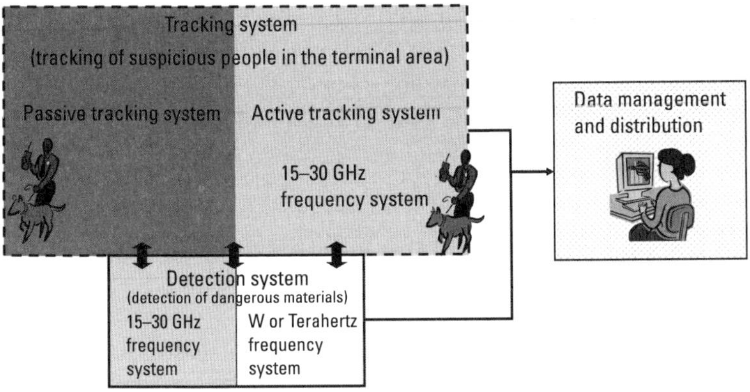

Figure 7.2 The structure of the distributed security system.

transit. This sensor system is required to detect and identify prohibited items (guns, knifes, nonmetallic weapons, explosives), reducing as much as possible the actions required to the passengers under surveillance. Among the possibilities for this sensor system is an advanced version of the millimeter-wave technology exploited inside the security scan; a brief review of detection sensors is provided in Section 7.4.

The second sensor system in the airport is equipped with new active and passive RF sensors that are able to track suspicious people (namely people who have been detected to potentially carry an illicit item), without interfering with passengers' transit and with other airport systems.

Finally, an airport security distribution network should be devised to integrate the data provided by these two sensor systems and allow security operators to face threats in the most suitable way, minimizing the risk for other people inside the terminal area.

Considering the security performance, it is interesting to notice that the presented distributed system has a number of advantages in terms of capability to provide privacy and a pleasant perception of the security process. The waiting time to provide screening is by definition not more of an issue, because the screening is performed continuously during the entire time that the passengers are inside the airport area, without requiring their cooperation. This is clearly in contrast with a specific screening time required at the security checkpoints where passengers should pass through. Therefore, the time response parameter does not apply to the new approach and only a processing effort is to be verified in terms of load on the security processing resources. This could mean that within the distributed system the waiting time and the throughput of passengers are nonrestrictive.

With reference to the other performance points, it is clear that the distributed detection, applied without cooperation, can require higher-quality sensors. However, the checking is allowed to be repeated many times during the passenger presence in the airport. Moreover, with this approach the security screening starts as soon as the passengers enter the airport area and not at the time of entrance inside the sterile area. This is a major advantage for the overall security of the entire airport area.

7.4 Sensors for a Distributed Dangerous Object Detection/Classification System

In this section an analysis of the requirements for the sensors to be used for the distributed detection/classification system is presented, together with a brief review of the technologies that are being considered for them.

In order to detect and classify dangerous and illicit objects of a small size, such as the ones reported in the list in [12], it is required to have a system with the following characteristics:

- Be totally safe for the passengers, as well as for the security personnel, so that their use does not cause any damage or issue for anyone's health;
- Provide at least 2-D (possibly 3-D) images to allow the use of the object shape as a basis for classification;
- A high geometric resolution, so that items of limited size can also be viewed;
- A good radiometric resolution, so that the classification based on clear images is reliable;
- A capability of penetration below clothes with a very low attenuation, so that the screening does not require to remove any of the clothing, possibly including coats, hats, and handbags;
- A strong response from the dangerous/illicit materials, including metallic, ceramic, and other nonmetallic harmful items, together with liquids;
- Possibly a response that is clearly dependent on the material in one of the available domains of analysis, as, for example, amplitude, frequency, or polarization, so that the object composition can be classified, thus increasing the overall object classification capability.

Moreover, the requirement to detect the illicit items in a distributed manner further requires that the sensors are able to operate:

- Effectively from a distance not limited to a few tenths of a centimeter, so that a limited number of sensors can be employed;

• Effectively for a broad range of illumination conditions and environmental temperatures.

As it is clear, these two additional requirements are desirable also in a concentrated detection/classification system, but assume a much higher importance when the detection is to be operated inside a distributed system.

Due to the significant efforts that are presently devoted to the improvement of the capability of such types of devices and to the resulting fast evolution of this technology, this section is largely far from a detailed and comprehensive state of the art on this subject. In fact, while a detailed state of the art would very soon be out of date, a quick overview of the categories of sensors that are being experimented gives a more useful and potentially slightly more persistent view of the types of sensors potentially usable in a distributed illicit/dangerous objects detection system. Therefore, we deliberately consider here a few categories of imaging sensors: passive millimeter-wave band (90–900 GHz) sensors, active millimeter-wave W-band sensors, active millimeter-wave K-band (10–35 GHz) sensors, and terahertz sensors.

Millimeter waves can easily penetrate clothing and hair, so they are especially suited for the detection/classification of weapons and explosives (and more generally of dangerous materials) hidden under people's clothing.

Passive sensors have the advantage of producing an image without emitting any electromagnetic (EM) radiation. This intrinsically provides two main advantages: passive sensors raise no issue about people's health or possible effects of radiation, and they do not need to use hardware and power to implement the emission of the exploited signals.

A passive sensor in the millimeter-wave regime is fundamentally based on measurement of the power received from the scene using a radiometer. The received radiation is the combination of the natural emissions from the body at the specific frequencies in use, with the reflection of the surrounding radiation from the human skin that has a reflectivity of about 10%. Its main detection mechanism is the following. An object concealed under the clothing blocks the natural emissions from the body, so that a change in the radiometric temperature is measured by the radiometer that is able to detect its presence if it has a geometric resolution high enough, compared to the object size. Very neat results are reported in [13], for both outdoor and indoor detection of metallic and nonmetallic weapons concealed under the clothes, by using a single 94-GHz radiometer mounted on a scanning 24-inch dish antenna adjusted to focus, respectively, at ~25 feet and at ~15 feet. The blocking mechanism applies to metallic objects and even nonmetallic objects, such as plastics and ceramics. Successful developments and detection experiments are also reported in [14], together with discussions on the classification capability of the sensors.

Other than blocking the natural emission of the human body, the objects reflect the radiation from the surroundings (in different ways and intensities depending on their material). As it is apparent, the EM radiation from the surroundings is largely dependent on the environment, showing different behaviors outdoors and indoors. Outdoors, the radiation from the sky is reflected off the skin of the upper part of the body, while the radiation from the ground (typically warmer) is reflected from the lower part. Indoors, assuming that a room is illuminating the body with radiation around ~295K, the body's warmer emissions tend to be blocked or attenuated, and the object can be revealed. However, since this additional reflection mechanism is not negligible with respect to the blocking, the detection/classification performance largely depends on the difference in the intensity of the radiation reflected from the skin and from the concealed object. This is easily calibrated in a very controlled indoor environment with assigned positions, while it can be very variable in outdoor or indoor conditions with large changes of position, illumination condition, temperature, and people density.

Active millimeter-wave sensors scan the environment with EM waves using single or distributed transmitters. Using known transmitted signals, they are capable to extract weak target responses from competing sources of noise. These devices are claimed to be able to provide quasi-3-D images by scanning around the body. The Security Scan sensors, briefly introduced in Section 7.2, are among the existing active millimeter-wave sensors already widely deployed at the airports for body scanning at the checkpoints. One of such types of sensors is the ProVision Body Scanning Checkpoint Security System from L-3 Communications Security and Detection Systems, Inc. [15]. The data acquisition and processing reduce the speed of passenger flow at a certain extent, but the technology is increasing quickly and the claims are for a throughput of 200–400 people per hour.

Very high resolution for the active millimeter-wave sensors is typically obtained either by using a very large array of receivers or by using a single receiver that is moved to collect signals backscattered from a person's body within a wide enough observation angle and applying the synthetic aperture radar (SAR) approach [16] or its equivalent inverse SAR solution (ISAR). Very high-resolution images with the neat image quality of vehicles and bicycles have been obtained by operating in the W-band of the millimeter-wave range (90–220 MHz), using ground-based SAR [17], which are the basis for new developments that promise improved capability also at longer ranges than inside checkpoint boxes. In this case the performance is reasonably independent of the external illumination/temperature condition, since the reflection of ad hoc generated signals is used by the radar. However, the capability to detect concealed items close to the human body is still an area that requires research and development and is expected to increase screening capabilities in the near future. In this direction is oriented the work referred to in [18], aiming at obtaining a new generation of screening sensors.

Achieving the high spatial resolution provided by the active W-band sensors requires the use of signals with a very wide bandwidth, which is increased by useful information provided by scanning a wider range of frequencies in terms of discrimination capability. Obtaining the same characteristics operating with the lower portion of the millimeter-wave band, namely the K-band, requires the use of a much higher fractional bandwidth, so that ultrawideband (UWB) radar sensors are used. The effort for the use of such sensors in the short-range screening (few meters), by exploiting the low peak power transmitters and the continuous wave operation, is providing appreciable results that might significantly contribute to the detection/classification of concealed weapons and dangerous objects [19]. In this case the use of the SAR technique to obtain high-resolution images is also quite essential and one of the problems to be solved for practical application is related to the long processing times required especially for the 3-D imaging applications.

The terahertz imaging technique is based on the use of the terahertz spectrum to detect concealed explosives, chemical/biological agents, and metal objects on the basis of their characteristic reflectivity in the frequency range. A terahertz reflection image of a person would show the outline of clothing and the reflection of objects beneath, such as concealed weapons. The skin of the body would appear dark in the image, thus preventing concerns of violating personal privacy. A number of examples of a passive 1.5-THz image of test scenes have been obtained by experimenters, where hidden weapons on human body are identified [20].

Interesting potentialities of the terahertz spectroscopy are also related to the capability to detect the presence of different types of materials. In particular, an important field of application seems to be the standoff detection, involving passive and active methods for sensing chemical and biological materials when the sensor is physically separated from the target. Although many techniques and methodologies have been developed and are in use for explosive screening when close proximity is possible, the screening from a distance of several meters is still an open problem that terahertz seems to allow solving [21]. While technology problems related with the generation, handling, and measurements in the terahertz band are still present, many progresses that look extremely promising for a medium-term future are occurring around the world for these types of sensors.

It may be easily argued that each type of sensor can either be applied as a single one, or interconnected to form a network of devices that increases both the spatial extent of the provided coverage, and the capability to detect and classify the dangerous objects. This is especially clear by considering that each one of the types of screening sensors introduced earlier is characterized by its own peculiarities, in terms of resolution, detection reliability, and classification potentiality, and only an appropriate fusion of the information from multiple sensors is expected to provide a viable, effective, and reliable solution for the open security problem.

7.5 Indoor/Outdoor Noncooperative People Tracking

The distributed security system, where the detection/classification of dangerous materials is not only performed in the close proximity of the checkpoints, implies that the suspicious person, carrying the potentially dangerous object, is not immediately stopped by the security people who are always present at the checkpoint. In contrast, there is the need for an alarm message to be handled to the closest possible security people who have to move and reach the potential terrorist. In the meantime, it is not reasonable to expect that this person remains in the same position, where the dangerous material has been detected, and instead he or she is expected to walk away so that the security people would be able to stop the right person only if an appropriate person tracking system is available. Moreover, an appropriate global security scheme must be in place, allowing all the appropriate procedures to be followed.

For initial implementations of the distributed concept, the detection sensors can be concentrated close to the access ways to the airport area, so that their range of operation should not require providing full coverage of the airport area and the requirements of the sensors can be slightly relaxed, without moving back to a totally checkpoint-based screening. This cannot be applied to the tracking portion, since a few tenths of seconds are sufficient for someone to move quite far from the detection position.

Among the challenging requirements for the person tracking subsystem, we point out the fully noncooperative characteristic of its operation. This is clearly necessary, since it is assumed that the first actions of a potential terrorist would include the sabotage of any type of cooperative transponder device assigned to a person to allow cooperative tracking. This characteristic excludes the use of radio frequency identification (RFID) devices as well as other radiolocalization systems based on the tracking of a receiver, such as GPS or cellular phone base tracking. Therefore, the tracking subsystem must be addressed by means of short-range radar sensors, developed to work in the indoor airport area and its surroundings. Two approaches can be exploited for this purpose: either the use of high-resolution millimeter-wave tracking radar sensors, or the exploitation of existing signals inside the airport area by developing an ad hoc passive radar sensor.

The active millimeter-wave-band radar must be characterized by a wide frequency bandwidth to allow the human targets to be discriminated and followed inside the scenario, a narrow antenna beam to provide the high cross-range resolution, and a fast scanning capability. Moreover, this radar should be provided by an adequate transmission power level to provide coverage of the whole airport interior. Obviously, a limitation in the emitted power level has to be guaranteed to ensure safe operation, despite the use of high-frequency radiation that is assumed to have a limited impact on a person's health. While not many such devices are available as products for human tracking purposes, some types of sensors with widely speaking similar characteristics have been recently developed for the automotive market. Therefore, the explored lines of research and development activities consist of attempting to adjust the characteristics of such sensors to the airport tracking needs. The main challenges of this concept are:

- Discrimination and tracking of people in a crowded environment;
- The robustness of the system against interference;
- The capability of the system of not interfering with other equipment;
- An EM field strength compliant with current EM exposure regulations;
- The use of advanced solutions based on a multistatic operation or array 3-D imaging to increase the discrimination of closely-spaced static or slowly moving human bodies.

An alternative possibility is to resort to the passive radar sensor. In recent years the use of passive bistatic radar (PBR) for surveillance purposes has received renewed interest [22, 23]. PBR exploits existing illuminators of opportunity to perform target detection, which results in the exciting possibility of low-cost surveillance, reduced pollution of the EM environment, and covert operation. Among all the emitters available in the environment, broadcast transmitters represent some of the most attractive choices for long surveillance applications, owing to their excellent coverage. Specifically, the most common signals for PBRs in use today are noncooperative frequency-modulated (FM) commercial radio stations, since they offer a good trade-off between performance and overall system development costs [24]. A number of studies have looked at the use of different analog signals such as HF radio and UHF television broadcasts [25–27] as well as digital transmissions such as DAB, DVB, and GSM [28–31]. In these applications, PBR is rapidly reaching a point of maturity. Basically, even though there are still many challenges to be solved, involving both technology development and processing techniques, PBR's practical feasibility for long-range surveillance purposes has been well established.

With reference to local area monitoring applications, aiming at the detection and localization of designated human beings within short ranges, wider bandwidth signals of opportunity should be exploited to achieve the required range resolution. To this purpose, wireless local area network transmissions might be considered. Wireless networking applications are proliferating at a very rapid rate for both commercial and private use and are typically largely available inside the airport area. One of the most popular systems consists of the IEEE 802.11 Standards-based (WiFi) technology [32], which offers reasonable bandwidth (range resolution) and coverage and wide accessibility. The WiFi router, used to provide coverage for an assigned area, might potentially act as an ideal illuminator of opportunity for short-range detection and surveillance using the PBR principle. The scheme of principle [Figure 7.3(a)] is obtained by collecting samples from both a transmitter of opportunity and a surveillance antenna steered toward the area that is searched for the presence of a target. The cross-correlation between the surveillance antenna signal and Doppler-shifted and delayed versions of the transmitted signal allows the production of the so-called 2-D cross-correlation function (2D-CCF), which allows the detection of potentially moving targets.

One of the advantageous features of this approach is that no extra signal is transmitted, which makes the sensor totally free from any issue related to human health. It also operates in a totally covert mode, so that the potential terrorist has no possibility to detect that he or she is being tracked.

However, WiFi transmissions are complex, a number of standards have been developed, and further versions are being investigated; the three most commonly being deployed are 802.11a, 802.11b, and 802.11g. According to the physical layer specifications adopted by each standard, in typical operation

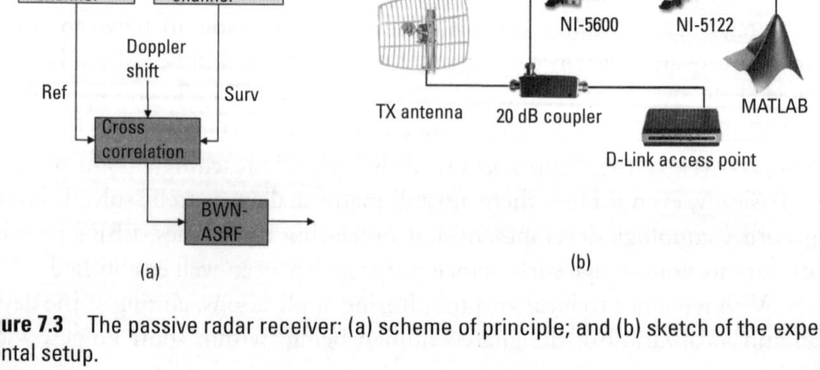

Figure 7.3 The passive radar receiver: (a) scheme of principle; and (b) sketch of the experimental setup.

modes, a WiFi access point (AP) may be broadcasting a mix of signals based on different modulations and coding schemes, each one characterized by its own autocorrelation function. A theoretical analysis of the WiFi signal ambiguity function has been presented in [33], with reference to different transmission standards. Moreover, preliminary experimental results have been reported in the literature, showing some interesting potentialities of a WiFi-based PBR.

Preliminary experimental results have been obtained within a research project carried on at the Infocom Department at the University of Rome La Sapienza. A dual-channel system prototype has been experimented, where two receivers collect phase-coherent digital samples, respectively, of the signal transmitted by the AP (via a directional coupler) and of the signal backscattered from the target and the background, as sketched in Figure 7.3(b), using National Instrument downconverters and digitizers.

As an example of the achievable results, an outdoor experiment is presented, where two targets were present (a car and a running man) moving forward the antenna location at approximately 25 km/h and 20 km/h, respectively, as sketched in Figure 7.4.

PBR operation inherently implies that the transmitted waveform is not within the control of the radar designer. In particular, following the IEEE 802.11 Standards, its ambiguity function is shown to provide a poor dynamic range. As is typical for PBR systems, the detection performance of a WiFi-based PBR may be strongly limited by the ambiguity function properties of the exploited signal. In fact, the target returns are likely to be masked by the high-level sidelobes corresponding to the ambiguity function of the disturbance

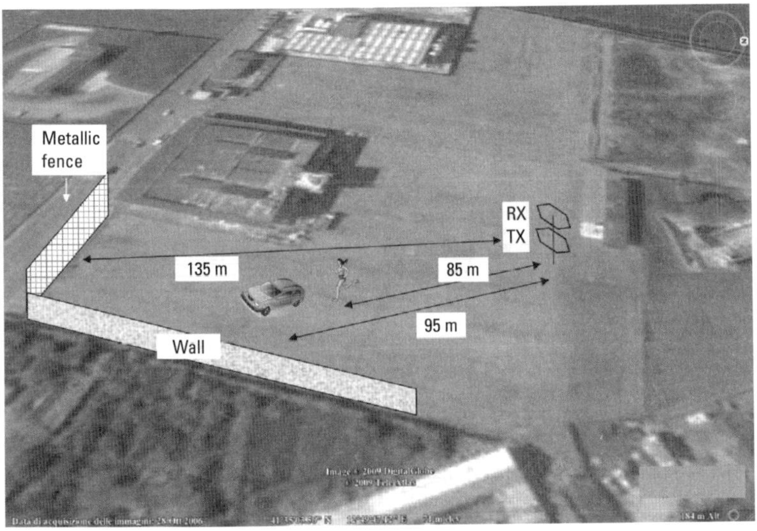

Figure 7.4 WiFi-based PBR experiment in Cisterna di Latina, Italy.

contributions (direct signal breakthrough and multipath reflections), even in the presence of a significant range-Doppler separation. However, proper data-processing techniques have been developed toward two different objectives: (1) to cope with the specific characteristics of the exploited waveform of opportunity, by suppressing its sidelobes, and (2) to obtain an effective removal of the disturbance contributions limiting the target detection [34, 35].

While the straight evaluation of the 2D-CCF provides the noisy image of Figure 7.5(a), where the two targets cannot be detected and tracked, after applying the developed signal processing technique, the results in Figure 7.5(b) are obtained, where now two spots appear at the range and velocity of the two targets present in the scene. This gives a preliminary confidence that some form of detection and tracking of human being in motion can be performed using a WiFi-based passive radar. However, some significant research activity still needs to be performed to assess whether the available resolution is enough for the required tracking characteristics and to make the tracking process robust, especially under changing conditions and a possibly crowded environment inside an airport area.

To exploit both the advantages of active millimeter-wave radar in terms of resolution and imaging capability and the attractive features of the passive radar, which still shows resolution and reliability limitations, an integrated system based on the fusion of the data from the two sensors is currently being conceived.

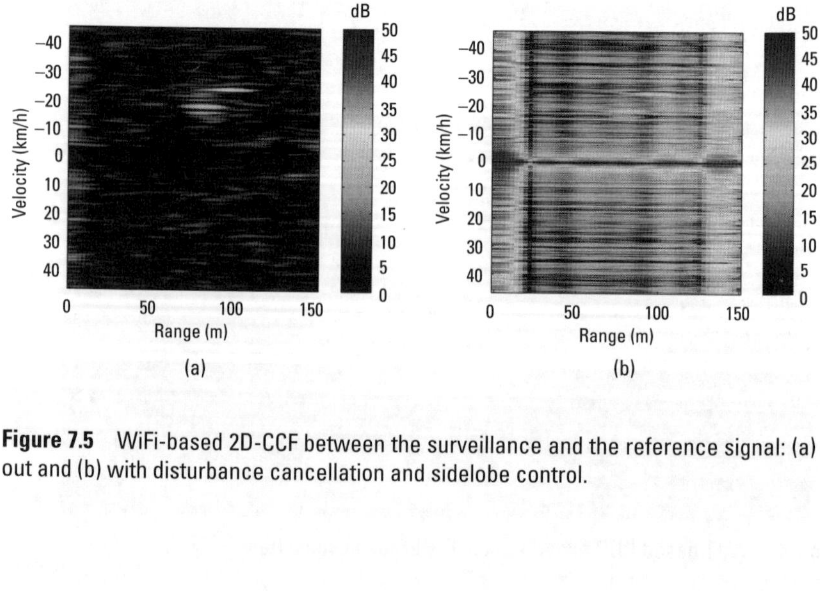

Figure 7.5 WiFi-based 2D-CCF between the surveillance and the reference signal: (a) without and (b) with disturbance cancellation and sidelobe control.

7.6 Conclusions

In this chapter the problem of air transport security was briefly introduced with a special reference to airport security. An innovative concept of distributed security was introduced as an alternative to the actual security system that is centered essentially in the location of the checkpoints that separates public and sterile areas. This approach has been shown to yield a number of advantages in terms of the capability to provide privacy and a pleasant perception of the security process.

The required technology for the implementation of such an approach was reviewed by discussing the increased requirements of the dangerous object detection/classification sensors, the need for a tracking subsystem, and a network operation for data fusion and data distribution. In particular, different types of sensors have been considered for the detection, including passive millimeter-wave, active W-band sensors, active K-band UWB sensors, and terahertz sensors, with their potentialities and problems. A possible tracking subsystem has also been introduced, showing that it can be based on both W-band active radar and passive radar exploiting the WiFi signals already present in the airport area. In particular, it has been shown that, at least from the preliminary experiments carried out with these systems, WiFi-based passive radars have the capability to effectively detect and track people without emitting EM radiation. Moreover, for both the dangerous object detection and the tracking subsystem, the need to capitalize the benefits of the different sensors by means of an appropriate data fusion scheme was mentioned.

As a final consideration, it is apparent that the innovative distributed approach does not necessarily imply the need to remove the present checkpoint-based security system. Namely, the presented approach can be potentially used to extend the coverage of the conventional security system instead of replacing it. Merging the two approaches, the surveillance starts when people enter the airport area, thus improving overall security. Obviously, this does not provide the gain of removing lines at the checkpoints; however, the lines might be greatly reduced on the basis that people arrive at the checkpoint after an initial screening already performed by the distributed security system.

Acknowledgments

The authors would like to mention the many fruitful conversations with Nicola Iarossi about the ATOM Project. They also acknowledge the collaboration of Paolo Falcone and Enrico Anniballi for parts of the described items.

References

[1] The FP7—AERONAUTICS and AIR TRANSPORT (AAT)—2008—RTD-1 ATOM Project Grant Agreement, https://atom-project.net/.

[2] http://www.rand.org/ise/projects/terrorismdatabase/.

[3] Frank, T., "Most Fake Bombs Missed by Screeners; 75% Not Detected at LAX, 60% at O'Hare," *The USA Today*, October 18, 2007.

[4] Wolff, S., "Cabin Baggage Screening: Best Practices and Effective Technologies," *Aviation Security International*, Vol. 16, No. 4, August 2010, pp. 16–21.

[5] Wolff, S., "Passenger Screening, Any Improvements?" *Aviation Security International*, Vol. 12, No. 8, February 2007, pp. 8–16.

[6] http://www.schiphol.nl/AtSchiphol/CheckinControl/InnovativeSecurityScanSimplifies-PassengerControl.html.

[7] Leo, J. G., and J. P. Lawler, "A Study of Passenger Perception and Sensitivity to Airport Backscatter X-Ray Technologies," *International Business & Economics Research Journal*, Vol. 6, No. 7, July 2007, pp. 11–18.

[8] Harris, D. H., "How to Really Improve Airport Security," *Ergonomics in Design*, Winter 2002, pp. 1–6.

[9] Sankaranarayanan, A. C., A. Veeraraghavan, and R. Chellappa, "Object Detection, Tracking and Recognition for Multiple Smart Cameras," *Proceedings of the IEEE*, Vol. 96, No. 10, October 2008, pp. 1606–1624.

[10] Elias, B., *Airport Passenger Screening: Background and Issues for Congress*, CRS Report for Congress, Congressional Research Service 7-5700, www.crs.gov, R40543, April 23, 2009.

[11] Elias, B., *Airport and Aviation Security: U.S. Policy and Strategy in the Age of Global Terrorism*, Chapter 7, Boca Raton, FL: Auerbach Publications, 2010, pp. 193–234.

[12] Regulation (EC) No 622/2003 amended by Regulation (EC) No 68/2004 and Regulation (EC) No 1546/2006.

[13] Yujiri, L., M. Shoucri, and P. Moffa, "Passive Millimeter Wave Imaging," *IEEE Microwave Magazine*, Vol. 4, No. 3, September 2003, pp. 39–50.

[14] Notel, D., et al., "A Demonstrator Approach for the Technology of a Passive Millimeterwave Imaging System and the Related Image Processing," *The Joint 30th International Conference on Infrared and Millimeter Waves and 13th International Conference on Terahertz Electronics, IRMMW-THz 2005*, 2005, pp. 319–320.

[15] http://www.dsxray.com.

[16] Franceschetti, G., and R. Lanari, *Synthetic Aperture Radar Processing*, Boca Raton, FL: CRC Press, 1999.

[17] Essen, H., et al., "High Resolution Millimetre Wave Measurement Radars for Ground Based SAR and ISAR Imaging," *IEEE Radar Conference, RADAR '08*, 2008.

[18] Hantscher, S., et al., "94 GHz Person Scanner with Circular Aperture as Part of a New Sensor Concept on Airports," *2010 11th International Radar Symposium, IRS 2010*, Vilnius (LT), June 15–17, 2010.

[19] Savelyev, T., et al., "Development of UWB Microwave Array Radar for Concealed Weapon Detection," *2010 11th International Radar Symposium (IRS)*, 2010.

[20] Luukanen, A., et al., "Passive Broadband Terahertz Camera for Stand-Off Concealed Threat Identification Using Superconducting Antenna-Coupled Microbolometers," *Proceedings of the 38th European Microwave Conference, 2008, EuMC 2008*, October 2008.

[21] Liu, H. -B., et al., "Terahertz Spectroscopy and Imaging for Defense and Security Applications," *Proceedings of the IEEE*, Vol. 95, No. 8, 2007, pp. 1514–1527.

[22] Special Issue on Passive Radar Systems, *IEE Proceedings on Radar, Sonar and Navigation*, Vol. 152, No. 3, June 2005, pp. 106–223.

[23] Lombardo, P., F. Colone, and A. Farina, "Passive Radar: Harvesting E.M. Radiations for Surveillance," *Tutorial Lecture, IEEE International Radar Conference*, Washington, D.C., May 2010.

[24] Howland, P. E., D. Maksimiuk, and G. Reitsma, "FM Radio Based Bistatic Radar," *IEE Proceedings on Radar, Sonar and Navigation*, Vol. 152, No. 3, June 2005, pp. 107–115.

[25] Griffiths, H. D., and N. R. Long, "Television Based Bistatic Radar," *IEE Proc. F, Commun. Radar Sig. Proc.*, Vol. 133, No. 7, 1986, pp. 649–657.

[26] Fabrizio, G., et al., "Adaptive Beam Forming for High Frequency Over-the-Horizon Passive Radar," *IET Radar, Sonar & Navigation*, Vol. 3, No. 4, August 2009, pp. 384–405.

[27] Thomas, J. M., H. D. Griffiths, and C. J. Baker, "Ambiguity Function Analysis of Digital Radio Mondiale Signals for HF Passive Bistatic Radar," *Electronics Letters*, Vol. 42, December 2006, pp. 1482–1483.

[28] Coleman, C., and H. Yardley, "Passive Bistatic Radar Based on Target Illuminations by Digital Audio Broadcasting," *IET Radar, Sonar and Navigation*, Vol. 2, No. 5, October 2008.

[29] Saini, R., and M. Cherniakov, "DTV Signal Ambiguity Function Analysis for Radar Application," *IEE Proceedings on Radar, Sonar and Navigation*, Vol. 152, No. 3, June 2005, pp. 133–142.

[30] Poullin, D., "Passive Detection Using Broadcasters (DAB, DVB) with CODFM Modulation," *IEE Proceedings on Radar, Sonar and Navigation*, Vol. 152, No. 3, June 2005, pp. 143–152.

[31] Tan, D., et al., "Passive Radar Using Global System for Mobile Communication Signal: Theory, Implementation and Measurements," *IEE Proceedings on Radar, Sonar and Navigation*, Vol. 152, No. 3, June 2005, pp. 116–123.

[32] IEEE Standard for Information Technology, "Part 11: Wireless LAN Medium Access Control (MAC) and Physical Layer (PHY) Specifications," IEEE Std. 802.11-1999, and following Supplements and Amendments (IEEE Stds 802.11a-1999, 802.11b-1999, 802.11b-1999/Cor 1-2001, and 802.11g-2003).

[33] Colone, F., et al., "Ambiguity Function Analysis of Wireless LAN Transmissions for Passive Radar," *IEEE Trans. on Aerospace and Electronic Systems*, Vol. 47, No. 1, January 2011, pp. 40–264.

[34] Falcone, P., et al., "Range Sidelobes Reduction Filters for WiFi-Based Passive Bistatic Radar," *European Conference on Radar, EURAD 2009*, Rome, Italy, September 30–October 2, 2009.

[35] Falcone, P., et al., "Experimental Results for OFDM WiFi-Based Passive Bistatic Radar," *RADAR 2010, IEEE International Radar Conference*, Washington, D.C., May 10–14, 2010.

Appendix 7A List of Airport Security Measure Failures with Worst Consequences

Among the worst consequences of the failure of airport security measures are:

- The terrorist attack popularly called the Lod Airport massacre (May 30, 1972), now known as the Ben Gurion International Airport, in Tel Aviv, undertaken by three members of the Japanese Red Army. Firing indiscriminately with automatic firearms and throwing grenades, they managed to kill 24 people and injure 78 others, before being neutralized (one of them through suicide).
- The bombing of Air-India Flight 182 (June 23, 1985), en route to New Delhi and Bombay via London Heathrow from Montreal. All 329 people on board were killed, which included 280 Canadian citizens and 136 children.
- The Rome airport attack (December 27, 1985) when four gunmen walked to the ticket counter of Israel's El Al Airlines and TWA Airlines, at Leonardo da Vinci Airport in Rome, and opened fire with assault rifles and grenades. They killed 16 people and wounded 99 others before three of them were killed.
- The Vienna airport attack (December 27, 1985) at the Schwechat Airport, where three terrorists threw hand grenades towards crowds of passengers queuing to check in for a flight to Tel Aviv. The terrorists killed two people and wounded 39 others.
- The bombing of Pan Am flight 103 (December 21, 1988), en route from London to New York, over Lockerbie, Scotland. All 259 people onboard were killed, as well as 11 people on the ground.
- The hijacking of Air France flight 8969 (December 24, 1994), en route to Paris from Algiers, by armed Algerian terrorists, who tried to blow up the airplane (laden with more than 20 sticks of dynamite) over Paris.
- The events of September 11, 2001 (the single deadliest airline catastrophe), when men armed with simple box cutters took over four passenger airplanes and deliberately flew with two of them into the New York World Trade Center. Both 110-story buildings were demolished within an hour of impact, killing more than 2,800 people. A third airplane was deliberately flown into the Pentagon in Washington, D.C.
- The Los Angeles Airport (LAX) attack (July 4, 2002), when Hesham Mohamed Hadayet, an Egyptian-born man, opened fire at the El Al Israeli Airlines ticket counter, killing two people and wounding three

others. Hadayet shot the El Al ticket agent and turned his weapon on passengers in line before being subdued by a guard and finally shot.

- The attempted bombing of the Rome offices of the Spanish airline Iberia (December 2002), when anarchists sent a highly explosive parcel bomb. This was the second parcel bomb in 24 hours sent to Spanish targets in Europe; the first was sent to the Barcelona office of El Pais. Both devices were books with pages removed and replaced with explosives, attached to a spring-action detonator. Neither device exploded thanks to careful monitoring and security.
- The attempted attack at London's Gatwick Airport (February 13, 2003), when Rahaman Alan Hazil Mohammed, from Venezuela, was arrested in London after authorities found a live grenade in his luggage, after he had flown on a British Airways flight from Caracas. The north terminal of Gatwick was closed while authorities disposed of the grenade. Mohammed is being held in a London jail for possession of a grenade for terrorist purposes.
- The attempted attack at Cagliari Airport, Italy (October 13, 2003), when a homemade explosive device, consisting of a camping gas cylinder and an electrically wired metal plate, was discovered and defused near a runway at the Elmas Airport, in Cagliari in Sardinia. Security officials noticed a light near a boundary fence at the airport and when they went to investigate, they found the device planted alongside the fence. The incident came only minutes after air personnel discovered that an Alitalia airliner had been tampered with and the airport was closed.
- The crash of the Sibir Airlines flight bound for Sochi (August 24, 2004), when about 500 miles south of Moscow, the plane dropped off the radar after sending a distress hijacking signal and all crew and passengers were killed. Several passengers who were checked onto the flight did not board the plane. Terrorism was considered the cause, supported by the fact that debris was spread over a wide area and by the fact that another Russian airliner had crashed only minutes before. Investigators announced that they had found explosives in the wreckage of both aircraft, supporting the hypothesis that terrorism was the cause of the blast.
- The bombing of the Madrid Airport parking garage (December 30, 2006), when a large van bomb detonated, killing two people and injuring 26 others. A caller representing the Basque Fatherland and Freedom (ETA) warned of the device prior to its detonation. The two men killed were both Ecuadorian immigrants who were sleeping in vehicles in the garage. This is the ETA's most deadly bombing since May 2003.

- The attack at the Glasgow International Airport (June 30, 2007), when two men drove a dark green Jeep Cherokee into the main terminal, causing the car to burst into flames but failing to detonate what would have been a suicide bomb explosion on the airport. Authorities believe that this attempted suicide bomb incident bears the hallmark of an al-Qaeda operation and is connected to the discovery of two car bombs in London.
- The attempted bombing of Northwest flight 253 (December 25, 2009), which originated in Amsterdam, the Netherlands, where Umar Farouk Abdulmutallab boarded the plane after beginning his journey in Lagos, Nigeria, and was headed for Detroit. As the plane descended towards Detroit, Abdulmutallab attempted to detonate a 80g of PETN explosives hidden in his underwear. The device capable of blowing a hole in an aircraft failed to detonate properly and only Abdulmutallab was injured. Abdulmutallab admitted to being an al-Qaeda operative who was trained for this attack in Yemen. Reports indicate the 23-year-old man was recruited into the organization as a student at University College London between 2005 and 2008. Al-Qaeda in the Arabian Peninsula claimed responsibility for the attack.

8

Privacy Versus Security: A Fight That May Turn into an Alliance

Adarosa Ruffini and Giorgio Franceschetti

8.1 Introduction

The form and function of privacy and security and safety, as permitted and prescribed values in the social system of reference, evolve with the culture, being of use to the necessities of the society as well as the individual. Indeed, the individuals know and become familiar with these concepts from birth, naturally assimilating them in the communities where they live. These concepts tend to become part of the individuals' personalities, albeit with different nuances depending on their levels of perception, and the concepts direct the individuals' behaviors and model their personalities. In this chapter we analyze the concepts of privacy and security and safety, broadening their conceptual perspectives.

The elaboration of this perspective stimulates and encourages the development of new intellectual paradigms that are added to the way that privacy and security have been handled until now. We verify how privacy and security could assume the characteristic of new shared models to be accepted by the international context, in order to homogenize the system of governance, as well as the financial, economic, political, technological, and informatics domains. This is particularly useful when these archetypes, which we call open models of guarantee, as well as safeguard measures, should be related to specific scenarios such as Homeland Security. There is no question that the latter has expanded its targets, including now not only the protection of citizens from terrorist and natural disasters threats, but also the security of infrastructures, such as telecommunications and transport, energy, and services provided in the finance and administration areas. These archetypes should be also related to intelligent surveillance of the borders of each state to counter, for example, immigration and illegal trafficking. We can even assume that an international outlook is developing, aimed to plan open cooperation for widespread participation and instatement of conditions deemed acceptable by all parties involved. This is even more evident when the globalization of scenarios and military operations prioritizes the needs of homeland defense and the protection activity becomes a primary object against either conventional or asymmetric threats.

A brief summary of the content of the chapter is now in order. The interaction between the two apparently contrasting requests of privacy and security and safety is discussed in Section 8.2. However, in order to avoid possible misunderstandings, Section 8.3 clarifies the distinction, if any, between the two words, security and safety, which are used as separate or combined entities, without any apparent rule. In Section 8.4 the reasons for a wider definition of the security scenario, as appropriate to our integrated system of living, are demonstrated. The conclusion is that privacy shares the same need of protection as security and safety, along the same lines of implementation, and this is consistent with the spirit of the title of this chapter. However, it is also noted that this mutual origin requires the establishment of new rules and common agreements in the frame of the international community. Examples of initial movements along this direction are detailed in Section 8.5. A presentation of a possible valuable solution to the problem is the contribution of Section 8.6.

Before entering the main part of the chapter, some few additional points are worth of presentation. There is no doubt that there is a large international debate on the privacy versus security topic, but this debate is essentially confined in the technical sector, namely, how the privacy access should be ensured while implementing appropriate security codes. As a matter of fact, this issue is largely present in other chapters of the book, for instance (but not only) in Appendix 1A, where financial, physical, and health infrastructures are deeply analyzed and technical procedures to save the privacy issue are presented. Obviously, properly designed and operating infrastructures are listed, where privacy is accounted for and deeply defended.[1] It is explicitly noted that these infrastructures are not just a technological exercise; more recently, their design has moved along a specific analytical model, the Helen Nissenbaum contextual

1. For exhaustive references on this issue, see Chapter 1.

integrity scheme,[2] but the basic philosophy has always been about the contrast between security and privacy.

The viewpoint of this chapter moves along a different path, preliminary to any concurrent system implementation: the privacy-security debate from an ethical, normative, and legal point of view is in order. As a matter of fact, the essential coincidence between privacy and safety is first discussed; then the security scenario is extended to include not only the person and his or her goods, but also his or her life behavior rights. Finally, it is suggested that privacy is a fundamental right that belongs to this scenario, thus to be included in the security issue. This statement should be obviously coded within the frame of an appropriate set of rules to be accepted at international level; the archetypal approach is just an exercise along these lines. In view of this discussion, the title of this chapter, as well as its content, is justified. Security and privacy should not be contrasting issues, so that a compromise must be reached at the level of their implementation. On the contrary, a wider perspective under the umbrella of unified international agreements might merge the two issues into a single one.

8.2 Privacy and Security and Safety: Their Definition and Correlation

It is a common definition that privacy represents a way of being of the person and consists of the exclusion of the knowledge and prohibition of indiscretion by others of private attributes of the person itself [1]. We might add that this way of being belongs to the broader category of ethical rights of individuals and translates their needs, of a purely spiritual order, to isolate themselves from others and from their social context of reference.

The subject of our discussion is when, how, and to which extent all this is deemed appropriate and legitimate. As anticipated in Section 8.1, this point is logically located before any technical exam of the privacy defense in the security and safety scenario.

Modern society poses the question of the increase in the means and opportunities that generate interference in the privacy of individuals. The protection of privacy, as an essential condition for the inclusion of citizens in our society of participation and information, is therefore the final acceptance of a major ethical-juridical principle. This must be understood both as a right to discretion, concerning safe intimacy in private and family life, as well as a right to the protection of personal data, which is presently the most innovative aspect.

2. A detailed presentation of privacy-minded health information systems is provided in Chapter 5.

In order to guarantee individual freedom, it is therefore necessary to give, receive, and exchange data within a comprehensive legislative coverage that should provide full respect to all the values of the human personality, including the new emerging ones in the context of our fast developing social organization. The need for self-protection is present in the genome of all biological forms and is, at the same time, a basic element of any type of human aggregation.

With this view in mind, the pursuit of security and safety, now a true science, has often been defined as the common impetus for achieving an organization of society that leaves as little space as possible to arbitrariness and chaotic behavior. The contemporary world has even developed a culture of security and safety, whose contents are now a key factor for the general welfare of the community, and for the efficiency and competitiveness of the economic and technological sectors. Even if the conceptual approach has sometimes been incoherent, the search and the formation of models, also legal, of great mobility and flexibility, aimed to develop new scenarios of safeguard, are a positive fact.

Usually, the term security and safety has been used to indicate objective conditions free of danger, or guaranteed against the occurrence of adverse events, with an obvious corollary: absolute security and safety cannot be translated in real life, because it would exist only in the complete absence of danger. Security and safety can be affirmed as a basic human right, as it permeates all aspects of life. It is necessary to the intrinsic existence of every human being; it is a fundamental part of all popular common feelings, of social life, of all activities in which homo faber (the active man) measures himself, from the private to the public level, from the national level to the international one.

As a result of this, it is possible to refer to security and safety in a variety of approaches, from philosophical to political, from military to diplomatic, and from economic-financial to technological. Each approach contributes to the concept and practice of security, with its own visions and notions, based very often on antagonistic assumptions. Perhaps we are still missing a general overview that could reconcile the various positions.

On the other hand, it should be noted that the development of the technology in favor of security and safety has led to forms of heavy erosion of privacy rights, for which the more obvious and alarming manifestations are hereafter listed:

- Broadening of some particular collections of information, as those related to telephone traffic data, electronic mail, Internet access;
- Government pretence to obtain information from any private person;
- Interconnection and access to public databases;
- Considerable pressure to remove some of the guarantees related to processing of personal data under the excuse of economic efficiency.

Therefore, it is necessary that the presence of a fundamental and sound legislative contribution acts as a protective measure in these situations.

Privacy and security and safety, deemed for these reasons as meaningful values of our civilization, must be guaranteed through the adoption of normative standards, prepared not only with regard to their immediate and formal usefulness, but rather to ensure and defend the basic primary rights of citizens. The natural instrument to guarantee these rights is the contract, the most fitting legal instrument of postindustrial society, to reach maximum degree of sharing and accord among the parties.

A fundamental point must be stressed. Those normative standards, implemented via appropriate contracts, should be developed within the global scenario that is now the common area of all the nations; otherwise, they will add to the already-existing confusion. In ancient times, our life was essentially confined in the *ἀγορα'* (agorà), an ancient Greek word indicative of the square of the village. The situation is now just the opposite: we live in a global village, and common, resilient rules of behavior must be agreed upon.

The complex relation of the public and private partnership should also be reconsidered. The traditional decisional process of the public power should, in fact, have in mind the interest of private actors. This could be realized through the creation of mixed panels and work tables that reinforce the participation of all interested private parties.

8.3 The Paradigm of Integrated Security and Safety Management

We can state that a balance is needed between two contrasting issues. From one side, the citizens' right to save their personal data must be assured; at the same time, protection of people and goods from risks must be accomplished. This fundamental problem has been categorized in terms of the duet of privacy and security and safety. To address the problem, the role of the two duet components must be clearly demonstrated. This issue is simple for the single term, privacy, but not transparent for the double term, security and safety. Therefore, it is preliminarily convenient to examine with some detail the meaning of the latter component of the duet, because the lexical distinction of the two words has no well-defined semantic basis.

Webster's Dictionary defines safety as "the condition of being safe: freedom from exposure to danger; exemption from hurt, injury, loss, or any other unwelcome event, as physical, social, spiritual, financial, political, emotional, vocational, psychological, educational." It may well then include the protection of persons or property.

Security is in turn identified as the condition of being protected from dangers, loss, or criminal acts. One can therefore conclude that security is a concept similar to safety, with the emphasis on protection from external threats and dangers.

Is there so much difference in reality? Some examples may be helpful to clarify the difference, if any.

The Berlin Air Safety Center (BASC), which began to operate immediately after the World War II, had as its main purpose to diplomatically ensure the safety of Allied Air Traffic, to and from the western sectors of Berlin, from the threat of the Soviet and East German air defenses.

- The goal of the Web magazine *Information Safety and Freedom* is to defend the freedom of the press from attacks and counterstatements all over the world.
- In the United States all workers must have a Social Security card (the Social Security number).
- A well-known shoe brand advertises a particular tracking boot with the slogan Security Boots.
- Examples, such as the listed ones, do not explain at all the difference in the meaning attributed to security and safety.

As a matter of fact, there is not a border distinctly separating security from safety, because being safe automatically implies existence and/or implementation of some kind of protection. For this reason it is more pertinent to refer to security and safety as a general condition of immunity from danger and define it unitarily, while stressing the different aspects that compose the immunity itself. We could simply refer to safety, specifying it in relation to its applications' context. Accordingly, we will use the single word security, instead of security and safety, from now on.

The need to offer a functional description of a context of security, of its fundamental components, and of its features and relations that govern it can be solved with the implementation of integrated security management. The adoption of this new paradigm entails:

1. On one hand, to relate security to its application contexts, and the individuation of well-defined secure areas such as:
    - International relations and politics;
    - Military (including terrorism, nonconventional conflicts, operations different from war or supporting peace) and legal issues;
    - The safety of workers, environmental and financial security, and similar issues.

2. On the other hand, to attribute specifically the responsibility of involved actors, and define the degree of their performances as well their involvement.

As a matter of fact, any valid organization or entity should have the tools to recognize, assess, control, or reduce any destabilizing event or risk that could cause injury and, in emergencies, to intervene to limit the damage and restore the status quo ante (previous state). This means, in short, that all types of risk should be considered, natural, human, and technological, intentional or accidental, external or internal, and against material or intangibles, goods, and people.

8.4 The New Interpretation on the Security Role

To well understand the matter we are presenting, it is advisable to make a basic preliminary consideration. Many changes, involving the true essence of people, are happening in the scenario we are considering.

Scanned through video surveillance and biometrics, people are pushed towards an environment that increasingly characterizes the individual as a networked person: a man perpetually on the network, configured to emit and receive pulses, which trace and reconstruct his moving, his habits, his relationships, thus altering the meaning and content of its self-determination. An emblematic example is the use of fingerprints as biometric data, mostly used for different purposes of identification, control, and access, thus transforming man into a living password. An example is given in Chapter 7, where the innovative airport screening system ATOM is presented and documented in the references. The system is aimed to assess the cleaning status of potential passengers from the security viewpoint, by exposure of all people inside the airport and not only of passengers at the security gates. The search is implemented via appropriate wireless sensors, and no information about the wireless search is in principle provided to the sensed people.

Therefore, the starting point of the analysis can only be based upon a different and more extensive reading of the concept of personal freedom, on which also the rights of privacy and security converge, in the actual global scenario. The key issue is the central value and free development of the human personality, with total mental and physical freedom, considered in its functional interdependence and absolute indivisibility. The freedom of the individual, so qualified, is the starting point of a number of other related rights as, for example, the right to his name, his image, and his distinctive marks, which expressly refer to his psychological integrity. However, psychophysical rights are also included, such as the right to his life and defense from attacks to his way of living, which imply concepts of privacy and security. In a single omni-comprehensive statement, this implies a declaration of the inviolability of personal freedom. In this context, values such as privacy and security can deploy all their synergistic effects, well beyond their traditional opposition due to the more thoughtful understanding of this common matrix, from which both originate.

This integration can be achieved creating cultural and social shared scenarios, as well as reciprocal and common guarantees that protect the individual from the physical viewpoint, as well as from intrusions into his privacy and also from partial or distorted representations of its personal identity. The regulative framework must be the effect of the balance between the self-interest to control data and the public interest to traffic and circulation of information. This would be possible by promoting a progressive growth of the activity of public functions adapted to new situations.

The enlargement of this system of standards to all states, to achieve not only a common protection against risks (technical, natural, and human), but also to build the right to control the use that others can make of information regarding the individual, can be a real platform of freedom and security internationally shared. A set of rules that wants to acquire a high degree of effectiveness and widespread acceptance must include:

- New rules that must adapt themselves to new situations;
- Simultaneous promotion of cooperation suitable to create support and international acceptance, all over the aspects of perceived and implemented security and privacy issues.

We should consider that the international community is always characterized, at the institutional level, by the independence and sovereign power of the component nations. Promoting the option of voluntary agreement, in a moment of such marked difference between the attribution of responsibilities between public and private actors, is an inevitable and sustainable consequence of a cultural choice made upstream, for which privacy and security are, however, the central theme of the development of primary rights of the individual and of the community. At the same time, it is necessary to establish control mechanisms, which can ensure that safeguards are respected, in order to observe the agreement and, above all, it is necessary that the operational procedures that implement these control mechanisms are actually applied.

We conclude as follows. Traditionally and as previously presented, security has been limited to protect the individual and his personal property from the physical viewpoint. If this protection is extended to forbid intrusions into his privacy and personal identity, then the security encompasses and covers his privacy rights, too. As already pointed out, this wider vision of security is a delicate and complex issue; to be truly agreed upon, it should be formalized in the frame of the international community.

8.5 The Regulatory Framework and the Open Model of Assurance

Constitutions, laws, directives, rules, and similar regulations are sources of right. By the expression sources of right, we mean the juridical facts or acts that affect the creation, the modification, or the abrogation of the regulations in force in a legal system. Each state has its own regulative function, so that every state has its own legal system.

The regulations of each state are binding and must be respected; in the case of no observance, there is the application of a sanction. These regulations are enacted according to a principle of effectiveness and structured in a hierarchical form; the superior one prevails over the inferior ones.

Also, the confederations, which are unions among states (for example, the European Union and the United States of America), have their binding regulations, which are accepted by all participating states. Because there is no institution that links all the states of the world, those states wishing to have shared reference regulations must adopt international conventions and willing regulations (for example, the ISO standard) to harmonize their rules

The Italian legal system, and the Code on Personal Data Protection, or Privacy Code [2], which repealed and amended the previous law, defines the regulatory framework, concerning the processing of personal data, within an organic system of protection of fundamental freedoms of the individual, including all electronic procedures and digital processing of personal data. The code expressly defines the gathered information, as well as its control and classification.

The protection of privacy was already considered in the Italian Constitution [3, Article 3, paragraph 2], and the right to protection of personal data is already clearly outlined in the European Convention on Human Rights (ECHR) [4]. The convention makes a fundamental distinction between the traditional respect for private and family life [2, Article 7] and the right to protection of personal data [3, Article 8]; the former is intended as the right to the intimacy of life privacy, whereas the latter is devoted to the informative privacy and represents its most innovative outline. As a matter of fact, ECHR makes it clear, binding to that effect all EU Member States, that the protection of the individual is implemented under two mechanisms: the first is in charge of the community, but the second one is under the responsibility and right of the individual—first of all, his consent. This means that, even if the concerned person cannot always exercise a priori (preliminary) control on data, he or she can still exercise his or her right to supervise the proper functioning of the circulation of information, and implement a posteriori (successive) verification of the truth of the data, enforcing his or her right of access, rectification, modification, and integration.

It is also noted the Directive 95/46/EC of the European Parliament and of the Council of 24 October 1995 [5] on the protection of individuals with regard to the processing of personal data, and on the free movement of such data. The directive requires that the laws of member states should protect personal information in both the private and public sectors. That legislation imposes provisions to block transfers of information to nonmember states that do not provide an adequate level of protection.

The guarantee entitled to the person, with respect to data processing, is enforced at the European Community level, relapsing then in all European states, through a complex regulatory system, which is based on:

- A series of general guidelines that require the implementation of the state-level regulatory principles discussed previously;
- A series of rules that specifically require certain obligations;
- A number of faculties referring directly to the concerned person;
- A range of tasks that the Guarantor Authority approved.

All this in order to make perfectly coincident the need of proper functioning of the system, excluding from it all data whose acquisition does not meet any superior interest, together with the need of protection of privacy, which can inject into the circuit only approved and corrected data.

The Italian Privacy Code [2] has well considered and accounted for the relevance of protection of personal identity, and it has also identified the hazards and the means of protection that might relate to what is called the fragmentation of the individual, as a consequence of information technology. The set of instruments prepared by the code, such as the required individual consent to the treatment of his or her data, the right to forbid data handling or enforce their cancellation, and the possible appeal for collective administration entrusted to the ombudsman and the courts, are thus safeguards that guarantee the trade-off between the individual and the community along political and social directions. This is a sort of anti-intrusion protection, which allows legal actions against individual or collective behaviors that might disturb, assault, threaten, or falsify the lives of individuals in the society.

It should be noted that the adoption of this legislative instrument, which we define as an open model of assurance, must be supported by an indisputable collective consensus. As a matter of fact, any legislative action concerning the security issue at large is essentially a compromise, a balance of power between the rights of personal freedom and the institutions' rights, and should evolve along the development and acceptance of the shared informatics scenarios.

The adoption of the code [2] as a legal instrument has been a winning choice. This so-called security package represents an open door to the future, because it is the starting point for the creation and constant updating of legislation in the security as well as related scenarios. The power conferred to the guarantor, allowing him or her to promote its subscription within the concerned categories, to ensure compliance with the mandatory laws of the state, to contribute to its adoption and compliance, enhances the value of this very strategic choice and the ability to offer even higher benefits in the future.

8.6 The New Model: The Archetypal Approach

There is an almost unanimous consensus to identify the safeguarding of the privacy, security, and safety with legal standard guidelines. However, there is still a long way to use legal reference models to implement a simplified, harmonized, and effective planning of privacy and security, in accordance with their functional dependence, both at the national and international levels. The different cultures within which the rules must be observed suggest the adoption of conventional agreements, so that the legal models could be extended to a larger number of nations involved in the agreement itself. It is convenient to adopt a new structured and dynamic model, an archetypal approach, which exceeds the limits of traditional models, which are intrinsically static and inflexible.

The origin of archetypal approach stems out from the theory of archetypal certification developed by Adarosa Ruffini [6]. A very brief presentation of its origin is now in order.

The archetypal approach is aimed to compose a new ISO standard, the Societal Security, Public Private Partnership: Guidelines to Set Up Partnership Agreements. At the international level, a draft standard on this issue was presented at the 2010 Tenth Meeting of ISO/TC 223, Societal Security, in Bangkok, Thailand. The draft standard has been examined and discussed in the frame of the Working Group 1, Framework Standard on Societal Security Management, and has been submitted by UNI (Italy) to the ISO TC (Technical Committee) 223 secretariat for NP (new project) ballot. In the future next meeting in Berlin, the NP will evolve in a WD (Working Draft) document. In Europe, the Commission CEN-TC-391, Societal and Citizen Security, decided during its 2009 Meeting in Bruxelles to improve a new work item, Modeling, along the lines of [6]. In Italy, the archetypal approach is currently being used to define rules and commitments that the stakeholders should follow in the frame of the Genoa port security plan.

Even after this succinct presentation, it appears that the archetypal approach is becoming the reference issue for a modern view of the privacy issue and is on its way to become a standard. Its methodology is hereafter presented.

The integrated privacy and security management system, as we articulate in their multiplicity aspects, involves a great number of players, each of them in charge of different functions and responsibilities and with their own procedures and equipment.

The first conceptual limit of the traditional methods is the level of the apical coordination, which only results as a collection of information and decisions, already taken within each jurisdiction, by a bottom-up procedure. It follows that the applied procedures are static and incomplete compared to the multifaceted potential case studies, especially under critical conditions, due to the inevitable structural or organizational changes that may occur over time. Another limitation, of a purely economic type, is to make compulsory modifications, in whole or in part, to the technical systems already used by different stakeholders, with significant expenditure to purchase new equipment.

An evolved paradigm to implement the relations between privacy and security is therefore characterized by the fact that both of them, at the highest level, should be standardized as a structured and coordinated set of procedures to dynamically enable their development, management, and interoperability. This is independent from the used technologies, in our case mainly software and computers, or the organizational and operational policies of reference. The new approach, which we defined as archetypal because it refers to conceptual models, is implemented as a top-down procedure, whose starting point is the sharing by the stakeholders of the principles listed hereafter.

- *Accountability:* The processes through which an organization makes a commitment to respond to and balance the needs of stakeholders in its decision-making processes and activities. This definition emphasizes the need for organizations to balance and prioritise their responses to different stakeholder groups, according to organizational missions and criteria such as influence, responsibility, and representation.
- *Fairness:* Closely related to good faith, which requires considering only common benefits and obliging the partners to carry on fair perform-ances. Good faith, as an ethical meaning, is a behavior that presumes to act correctly (i.e., not detrimental in any way against another person). This duty is not honored not only when a partner acts to produce intentional injury to the other one, but even if his or her behavior is not marked by diligent fairness and sense of social solidarity. The duty of the partner is to preserve the interests of the other, because everyone must cooperate to

achieve the common advantage. Obviously, the partners should comply with national laws, professional ethics, and local regulations. Therefore, the rules and procedures should take into account the specific laws and regulations of each partner; the common task is to pursue the benefit of the society.

- *Transparency:* The provision of accessible and timely information to stakeholders. To be transparent, the organization should report to stakeholders on its activities and performance, providing basic information about what it is doing, through reports, information, and performance evaluations. Protection of sensitive information covered by privacy must be assured.
- *Participation:* This implies that stakeholders play an active role in decision making, protecting their right to privacy, but graduating their priorities on behalf of the highest right of collective security.

Implementation of this model requires a number of steps that are hereafter indicated.

1. The first step is to build some Cartesian grids, in which the *x*-axis corresponds to the privacy and security targets, and the *y*-axis corresponds to the activities to be undertaken by partners, in order to settle on the tasks that each partner must perform, with an indication of its corresponding responsibilities.
2. The second step, bearing in mind the basic principles listed earlier and any constraints due to mandatory standards, is to proceed to define the rules of relationships among the partners, the dynamics of functional interoperability of the system, the process of governance, and auditing.
3. The third step is to develop procedures and protocols necessary to meet the system objectives.
4. The fourth step is to proceed to the last certification.

The application of this new model for the synergistic management of privacy and security is an essentially innovative architecture, capable of identifying the complex system of relations among actors and forced to comply with the basic principles that allow:

- Transparency and sharing of procedures;

- Incontrovertible identification of the levels of decision and responsibility for any action;
- Maximum degree of coordination of subsystems activated during the critical event, while respecting the internal procedures of each actor;
- Dynamic adaptation of the procedures depending on the type of event, on the time and operational context, and on the priorities for action;
- Flexibility, to adapt the protocols to mandatory regulations, national and supranational.

A pictorial representation of this top-down process, leading to the archetypal configuration of the process, is depicted in Figure 8.1, which briefly summarizes the three levels leading to put into operation the archetypal approach. At the top level the interaction among models, standard requirements, and needed procedures is shown, leading to the novel archetypal approach. The intermediate level shows the implementation phase, when processing and

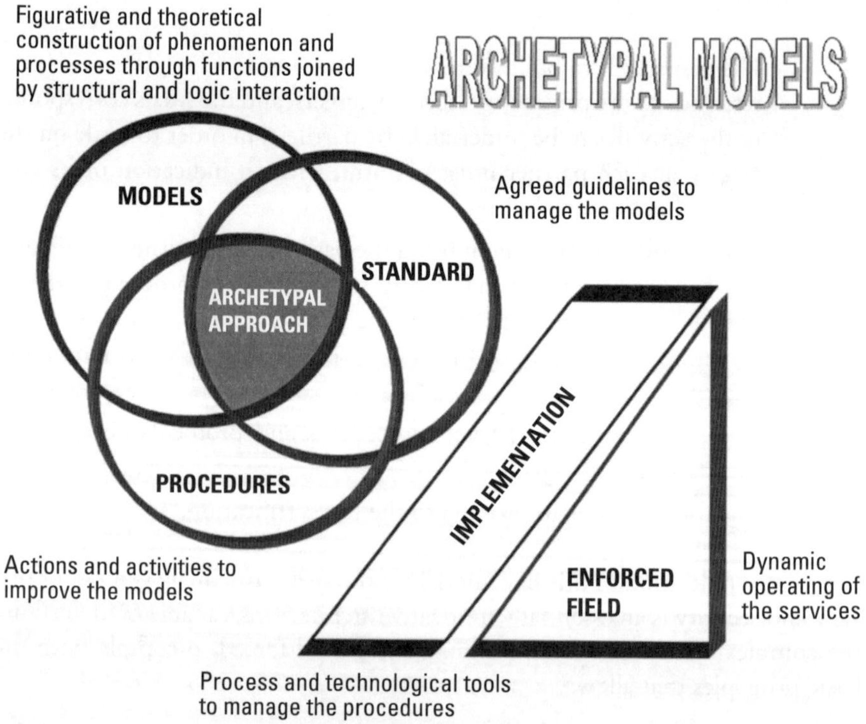

Figure 8.1 The three structural components of the archetypal approach: models, agreed guidelines, and procedures to generate the approach; transition to its real implementation; and its enforcement over the operational field.

technological tools allow the transition from a conceptual approach to an operating system. The third level represents the operative experimentation and final application.

8.7 Conclusions

Our discussion, albeit briefly argued, showed at a glance a problem of great legal significance: the formation of norms that are most suitable to govern the phenomena of great significance and importance such as the privacy and security and safety. The first step has been the elimination of the double reference security and safety, in favor of the single word security, due to the essential superposition of the two words, security and safety, in the Homeland Security scenario. Then we sought to offer insights and forms of resolution to the apparent contrasting issue between safety and security, trying to overcome this apparent opposition, and to achieve an alliance. The simple and transparent concept is that security is aimed at the protection of the individual as a person, as well as the protection of the individual as a person, as well as of his or her tangible and intangible goods. If the privacy, as fundamental right of any person, is included among the latter, then this opposition becomes an alliance. Obviously, the implementation of this simple statement faces a large number of problems, which have been discussed and analyzed.

The protection of the free development of personality, individually and/or collectively, is a core value of our current context of globalization. In conclusion, a possible line of solution, the archetypal approach, has been presented.

According to the parameters that we have shown, the conventional forms of discipline and covenantal agreement are the only ones able to acquire international importance. These forms, if properly used, help to create a conceptual framework of harmonized protection that is shared by as many states as possible. In this way we find relevant and effective responses to real needs of privacy and security, for their extensibility to all mankind. Maladministration or failure risks are minimized, according to the proactive commitment to accountability by public and private operators in charge of the shared framework of protection.

References

[1] Barth, A., et al., "Privacy and Contextual Integrity: Framework and Applications," *Proc. 27th IEEE Security and Privacy Symposium*, May 2007.

[2] Law of 31 December 1966, No. 675, Code on Personal Data Protection, or Privacy Code (Law 31 December 1996, no. 675, Delega al Governo in materia di tutela delle persone e di altri soggetti rispetto al trattamento dei dati personali).

[3] Italian Constitution, La Costituzione della Repubblica Italiana, December 27, 1947.

[4] European Convention on Human Rights (ECHR): Convention for the Protection of Human Rights and Fundamental Freedoms, September 3, 1953.

[5] http://www.caslon.com.au/privacyguide4.htm (privacy legislation, reports and developments in the European Union).

[6] Ruffini, A., *Certificazione e sicurezza: Genesi e sviluppo di un modello culturale*, Rome, Italy: Franco Angeli Publisher, 2007.

9

Privacy-Aware Design for the Monitoring, Control, and Protection of Critical Infrastructure

Stephen B. Wicker

9.1 Introduction

In the past decade, sensor networking technology has made dramatic advances. Sensing systems such as the Berkeley/Intel mote have become increasingly small and powerful, while emerging applications are as diverse as home health care, defense against bioterrorism, and the control of smart grids. As our ability to collect intelligence from structures and the environment has improved, so has our ability to use this data to monitor, control, and protect critical infrastructure. The resulting improved efficiency reliability will have a dramatic impact on the world economy. However, while the use of sensing systems offers powerful and long-term support for critical infrastructure, a problem lies in their use for data acquisition and control at ever decreasing levels of granularity. As we show in this chapter, sensor data can become determinative at a personal level, providing, for example, personal details about an individual's behavior within the home through insights into the individual's use of water, heat, and electricity. Technical solutions may thus evolve into privacy problems, which can, in turn, become legal problems for the utilities, corporations, and agencies that depend on sensor data collection.

This chapter begins by illustrating the powerful potential for sensing systems in protecting critical infrastructure. It then discusses the nature of privacy and the emerging concerns created by advanced sensing systems: a solution in the form of a set of principles for privacy-aware design is introduced. A case study focusing on demand response systems is developed. It is shown that the fine-grained power consumption data needed to implement demand response presents a serious privacy problem, as this data reveals behavior within the home; a privacy-aware design is then presented that implements demand response while respecting the privacy of consumers.

9.2 Sensor Networking Technologies for Infrastructure Protection

As sensing system technology has developed over the past decade, potential applications have expanded dramatically. One area of focus has been on embedded sensor networks for critical infrastructure [1]. Critical infrastructure applications include the power grid in general and supervisory control and data acquisition (SCADA) systems in particular, transportation, and water and fuel distribution. The increased emphasis on these applications has been driven in part by the development of supporting technology for randomly distributed, wireless sensors [2]. Such sensors greatly ease deployment in buildings, allowing for simultaneous surveillance and energy management in existing structures, as well as incorporation into building materials in future structures. Other relatively low-cost applications include security for open spaces such as parks, plazas, and street corners. Such deployment allows for the combination of surveillance with environmental monitoring.

Novel developments in the sensors themselves allow for a wider variety of public applications. For example, researchers at Cornell University and the Wadsworth Laboratory of the New York State Department of Health have developed a biosensor based on a bilayer lipid membrane [3]. As seen in Figure 9.1, a protein (the object at the center of the figure) is inserted into the membrane to create an ion channel. The cis compartment denotes the location of the material to be tested, while the trans compartment is the far side. When a target toxin in the cis compartment binds with the protein, the conductivity of the ion channel increases, creating an increased current flow that can be detected by the surrounding electronics. The target toxin can be varied by selecting different proteins. Such biosensors offer great potential for protecting water supplies and large-scale air conditioning units against the accidental and/or intentional introductions of hazardous materials, such as *E. coli* bacteria.

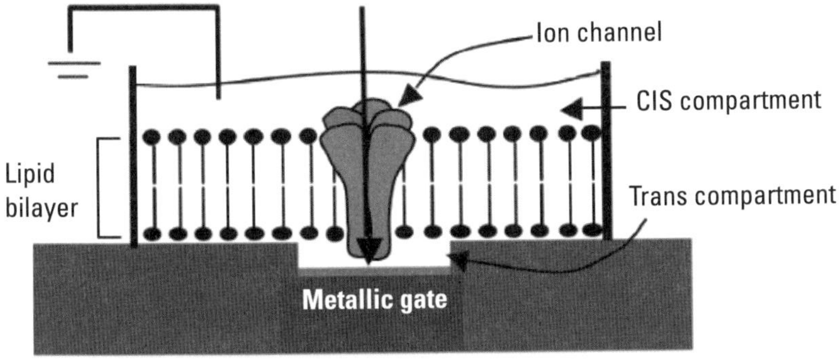

Figure 9.1 Designer biosensor based on bilayer lipid membrane.

Recent developments in long-term power sources have increased the potential for embedding sensors in critical infrastructure. Such power sources allow for long-term deployment in inaccessible remote facilities and in building materials. Researchers at Cornell University have focused on batteries that are powered by low-level radiation [4]. A variety of radioisotopes have been shown to have the potential for powering batteries with a 50-year life and a continuous power density of 1–10 mw/cm^2. As seen in Figure 9.2, two pieces of differently doped semiconductor are joined together. The two pieces are bombarded with beta particles from a radioisotope, freeing ions in the semiconductor materials. As the two materials are differently doped, there will be a charge differential in the freed ions, thus creating an electromotive force.

A variety of SiC-based beta-voltaic cells have been developed and tested. Thus far, the best measured power density for a Ni-63 source is 5.6 nW/cm^2 with a 4.4% efficiency. A tritium source provided ~1 nW/cm^2 with a 10%

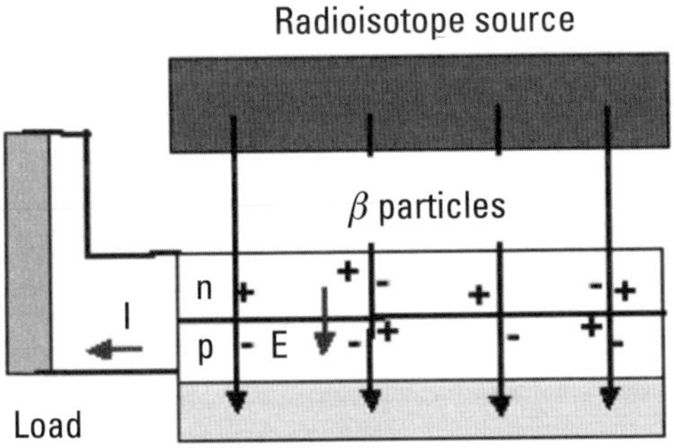

Figure 9.2 Beta-particle generated EMF.

efficiency. Target applications, such as that depicted in Figure 9.3, include long-term deployment in urban infrastructure.

There have also been significant developments in the electronics that support the sensor on the sensing platform. Researchers at Cornell University, working within the TRUST NSF Science and Technology Center, have developed an asynchronous processor that uses event-driven execution to dramatically lower power consumption [5]. The combination of event-driven execution and low-power consumption is ideal for sensor platforms that are intended to provide alerts when toxin levels exceed a threshold. The logic in this asynchronous processor has no clock, thus eliminating spurious signal transitions. As the hardware is only active when computation is needed, performance on the level of 24 pJ/ins and 28 MIPS at 0.6V is possible. Figure 9.4 depicts the organization of the processor. Comparative performance statistics are provided in Table 9.1.

As can be seen from Figure 9.4, multiple instructions can be fetched from instruction memory (IMEM), decoded, and passed to execution modules, where they will reside until events indicate that execution is appropriate. The timer controls the fetch and decode processes, but not the execution process, which is asynchronous.

9.3 Example Application Development: SCADA

Next Generation Supervisory Control and Data Acquisition technology (NG-SCADA, henceforth, SCADA) holds the potential for greatly increasing the

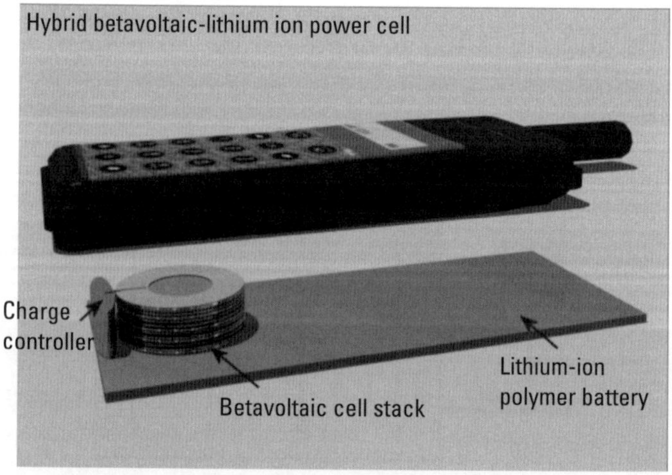

Figure 9.3 Beta-voltaic power cell intended for a cellular phone.

Table 9.1 Comparative Performance Statistics for the Cornell Processor

Processor	Bus	Year	Energy/op	Operations/sec
Atmel	8	2008	1–4 nJ	4 MIPS
StrongARM	32	2002	1.9 nJ	130 MIPS
MiniMIPS	32	1998	2.3 nJ	22 MIPS
Amulet3i	32	2000	1.6 nJ	80 MIPS
80C51 (P)	8	1998	1 nJ	4 MIPS
Lutonium	8	2003	43 pJ	4 MIPS
Cornell Processor	16	2003	24 pJ	28 MIPS

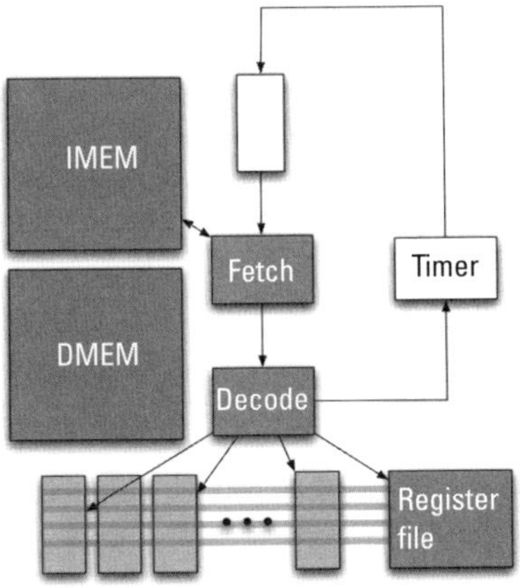

Figure 9.4 Cornell asynchronous processor architecture.

flexibility and efficiency of power grids and other delivery systems and industrial processes across the world. SCADA is the system that collects information from a piece of physical infrastructure, processes that data, and provides it to a central collection point. SCADA also provides a return path, allowing for operation of actuators that control the behavior of physical infrastructure. It is important to note that, while SCADA is generally taken to refer to monitoring and control of the power grid, it has more general application to water, gas, and manufacturing plants.

An example of SCADA architecture is shown in Figure 9.5 [6]. Starting at the bottom of the figure, the physical infrastructure, which can take many forms, is monitored and manipulated through a series of sensors and actuators. The sensors and actuators are in turn monitored and controlled by remote terminal units (RTUs). An RTU is typically a small computer designed for remote deployment and distributed processing. The RTUs are typically connected to

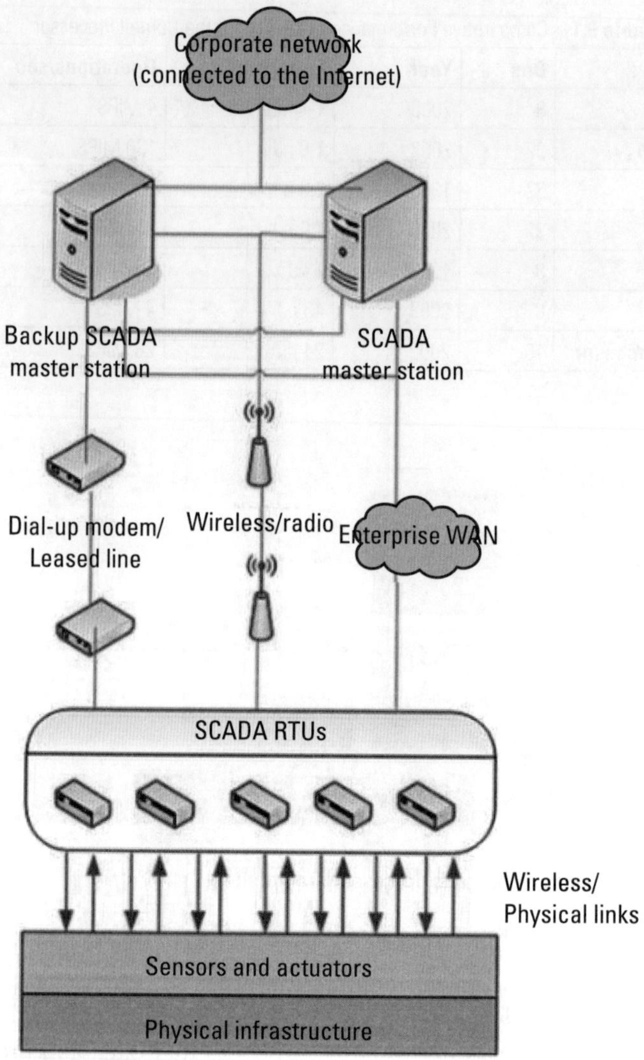

Figure 9.5 The SCADA architecture.

the SCADA master station through wired or wireless networks. The master provides an interface to human operators, as well as providing centralized automatic control functions.

To facilitate the design and testing process, TRUST researchers at Vanderbilt and Berkeley have developed a simulation testbed aimed to assess vulnerabilities of current SCADA implementations in realistic settings and to test solutions to address such vulnerabilities. The testbed also allows researchers to experiment with innovative architectural and technological solutions for next generation wireless SCADA. An open-source design has been adopted to create

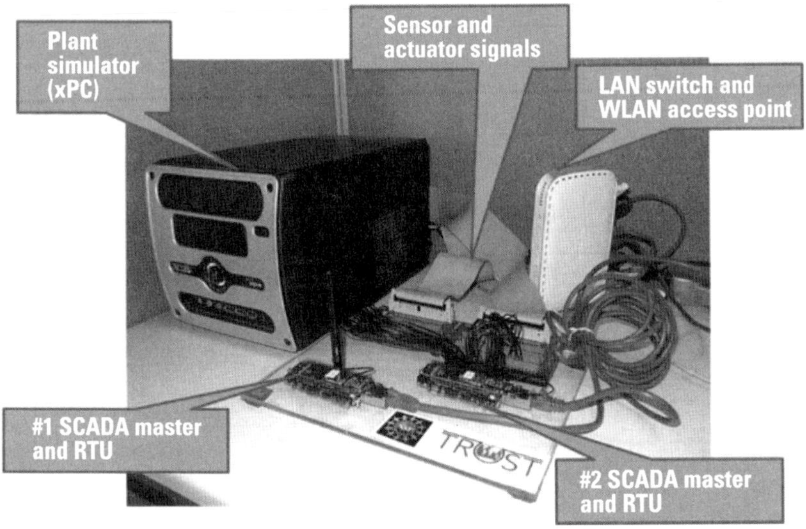

Figure 9.6 The TRUST SCADA testbed.

an affordable and highly flexible testbed for the TRUST community and for national needs through the Department of Energy laboratories Sandia and Oakridge.

An early version of the testbed is shown in Figure 9.6 [6]. A plant simulator was created using a Simulink software running on an xPC hardware platform. The RTUs were emulated using Robostix microcontrollers, while the SCADA masters are implemented using Gumstix Linux processors. The plant, RTU, and master processors are independent, allowing for a simulation environment that can emulate the reality of a remote sensing and control system.

Figure 9.7 depicts a chemical plant that can be simulated on the xPC target machine. This figure shows an adaptation of a publically available chemical plant model that has 4 processes, 16 control loops, 12 control variables (actuators), and 8 measured variables (sensors). In a simulation of this chemical plant, TRUST researchers are able to simulate 1 hour of operation of the plant in 1 minute, allowing for realistic simulations over a variety of timescales. A recent set of experiments focused on plant behavior under a variety of networked attacks. Researchers were able to show that a hacked control loop could be used to rapidly push the plant into a danger zone. By identifying such vulnerabilities, it is then possible to address them through redesign or additional detection and control mechanisms.

In subsequent sections it is shown that advanced sensing systems such these create substantial privacy problems. Before making this assertion, it is first necessary to define privacy, and then assess the extent to which privacy is protected by the legal system.

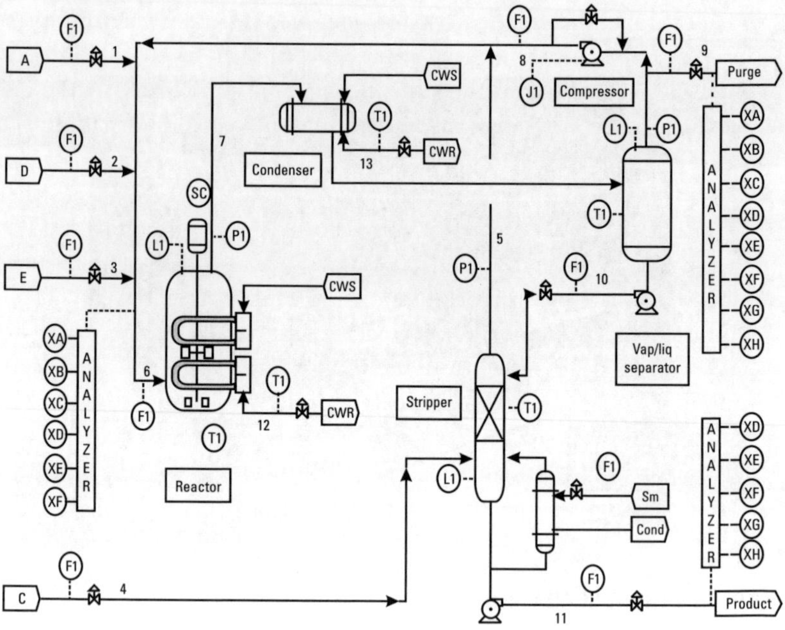

Figure 9.7 The Tennessee-Eastman chemical plant. (From: [7].)

9.4 Privacy: Definitions and the Law

The impact of new technologies on privacy is actually an old problem. In 1890 the *Harvard Law Review* published an article by Samuel Warren and Louis Brandeis entitled "The Right to Privacy." This article, often proclaimed the most influential law review article ever written, identified a right to privacy in existing law while decrying the impact of novel technologies, in this case, the "instantaneous photographs" [8]:

> Instantaneous photographs and newspaper enterprise have invaded the sacred precincts of private and domestic life; and numerous mechanical devices threaten to make good the prediction that "what is whispered in the closet shall be proclaimed from the housetops."

As is shown in the next section, residential power monitoring, while a useful tool in the development of demand response systems, may constitute a significant invasion of "the sacred precincts of private and domestic life" in the current regulatory environment.

To understand the impact of monitoring, it is necessary to have a working definition for privacy. Privacy has often been characterized in terms of the right to be let alone; this frequently cited reference is from Cooley on torts [9], and is quoted, for example, in the Warren and Brandeis article [8]. Subsequent commentators have added further dimensions and complexity to the definition by focusing on control over the revelation of personal information. Erving Goffman's *The Presentation of Self in Everyday Life* is a key reference that focuses on the definition of self in terms of selective disclosure [10]. Others further refine this definition by viewing privacy in terms of personal dignity and liberty. Building on the work of Bentham [11] and Foucault [12], scholars such as Julie Cohen have recognized that questions of informational privacy implicate self-definition, perceptions of public spaces, and the foundation of a liberal democratic state [13]. It follows from such work that when power consumption is monitored at a level at which personally identifying information is collected, issues of behavior modification and personal liberty are implicated.

It is possible and perhaps likely that the utilities will use collected information for purposes other than those stated. Oscar Gandy, Daniel Solove, and others have noted the alacrity with which corporations convert personal information that is collected in the course of business into a commodity and sell it, often without any acknowledgment of the rights of the parties whose personal information is being sold [14, 15]. State agencies are not immune from this tendency; most notably, State Departments of motor vehicles have sold information from driving records to individuals and commercial firms. Congress legislated against this practice in 1994 by passing the Driver's Privacy Protection Act (DPPA).[1]

Assuming that the collection of power consumption data implicates privacy (we show in Section 9.7 that this is, in fact, the case), the question then arises as to whether federal law provides some protection. In *Katz v. United States* (1967), the U.S. Supreme Court ruled that Fourth Amendment protections against unreasonable search and seizure covered conversations by individuals in a telephone booth. In a concurring opinion, Justice Harlan formulated a two-part test in which a search is determined to have occurred when the individual: (1) has exhibited an actual (subjective) expectation of privacy, and (2) society is prepared to recognize that this expectation is (objectively) reasonable.

The question of an expectation of privacy became the deciding (and negating) element in subsequent decisions involving data held by third parties. In *Smith v. Maryland* (1979), the U.S. Supreme Court held that the capturing and recording of dialed telephone numbers did not constitute a search, as any user of a telephone must be aware that they are conveying dialed numbers to a third party, namely, the telephone company. Similarly, in *United States v. Miller*, the

1. The passage of the act was in part motivated by a tragic incident. A California DMV employee provided information about an actress, Rebecca Schaeffer, to a private investigator. The private investigator then gave the information to a stalker, who proceeded to murder the actress. See, for example, Solove, Rotenberg, and Schwartz, *Information Privacy Law*, 2006, p. 618.

U.S. Supreme Court held that cancelled checks were the business records of the bank, and that the bank's customers had no reasonable expectation of privacy in the date reflected in those records.

Protections against actual physical intrusion into the home are still in place, however, and it would seem that the use of data-collecting technologies that are somehow equivalent to an intrusion are also forbidden without a warrant. In *Kyllo v. United States* (2001), the U.S. Supreme Court held in a 5–4 ruling that the thermal imaging of a suspect's home from outside that home constituted a search. Justice Scalia, writing for the majority, sensing a physical intrusion:

> Where, as here, the Government uses a device that is not in general public use, to explore details of the home that would previously have been unknowable without physical intrusion, the surveillance is a "search" and is presumptively unreasonable without a warrant.

The tenuous status of this understanding was reflected in Justice Stevens' dissent. He seemed to revisit the same third party, outside the home arguments seen in Smith and Miller.[2] Stevens focused on the primitive nature of the technology in sanctioning its use without a warrant:

> The Court has crafted a rule that purports to deal with direct observations of the inside of the home, but the case before us merely involves indirect deductions from "off-the-wall" surveillance, that is, observations of the exterior of the home. Those observations were made with a fairly primitive thermal imager that gathered data exposed on the outside of petitioner's home but did not invade any constitutionally protected interest in privacy.

At least on its surface, it seems that power consumption data falls into the same constitutional bin as dialed numbers and checks, information freely given to a third party. The author does not agree with this position, but it stands as the basis for a first principle of information privacy: if one wishes to regard information as private, then one should not let it pass through the walls of one's home. In this chapter we explore means for quantifying and limiting the extent to which personally identifying information passes across the boundaries of the home, establishing tools for evaluating threats to individual privacy and autonomy.

---

2. The Olmstead majority opinion that sanctioned wiretapping without warrants is based in part on the tapping taking place outside the home (277 U.S. 438 (1928)). Note also the now-famous dissenting opinion in Olmstead by Justice Brandeis, who 38 years earlier had coauthored "The Right to Privacy," as discussed earlier.

9.5 Privacy Issue: The Market for Information

In evaluating the privacy concerns created by sensing systems, the focus is often placed on third-party surveillance (hackers, eavesdroppers, and so forth). Though a very real problem, this is not our focus here; instead, we consider the arguably more pervasive privacy problems created by the collection of user information by the service provider. As we show for the specific case of power consumption data, this information can often be repurposed to provide information about the behavior and inclinations of the user, information that is of great values to marketers and others. The tension created by potential financial gain on one hand and the harm to the consumer through potential surveillance and manipulation on the other imply that the collection of such personal data results in a moral hazard for the service provider.

In *The Panoptic Sort*, Oscar Gandy's seminal work on the collection of consumer information, it is shown that the goal of information collection is discrimination [14]. Consumers are partitioned into groups that are more or less likely to buy given products. The consumers are then relegated to different information streams according to their perceived interest in and ability to buy a product. The immediate goal of the marketer in creating this vast consumer sorting information is efficiency: highly-focused marketing strategies make more money and are thus a better use of advertising dollars. The long-term result, however, is that various buying groups view different information streams, and some are disadvantaged accordingly.

This is a growing problem. According to Gandy, in a 1996 Gallup poll it was found that 77% of U.S. companies use some form of direct mail, targeted e-mail, or telemarketing. In 2003 telemarketing was a $662-billion-per-year industry. Today direct marketing is an enormous industry. According to the Direct Marketing Association, $149.3 billion was spent on direct marketing in 2009—more than half of all advertising expenditures in the United States—with a return of close to $1.783 trillion in sales attributable to the advertising.[3] The latter amounted to 8.3% of total U.S. gross domestic product.

3. In 2009, marketers—commercial and nonprofit—spent $149.3 billion on direct marketing, which accounted for 54.3% of all ad expenditures in the United States. Measured against total U.S. sales, these advertising expenditures generated approximately $1.783 trillion in incremental sales. In 2009, direct marketing accounted for 8.3% of total U.S. gross domestic product. Also in 2009, there were 1.4 million direct marketing employees in the United States. Their collective sales efforts directly supported 8.4 million other jobs, accounting for a total of 9.9 million U.S. jobs (http://www.thedma.org/aboutdma/whatisthedma.shtml).

9.6 Protecting Privacy: Privacy-Aware Design

In this section we summarize a framework, developed in [16], for privacy-aware design. The framework consists of a set of principles that were derived from the Fair Information Practices proposed by the Department of Health, Education, and Welfare (HEW) in a 1973 study entitled *Records, Computers, and the Rights of Citizen* [17]. The framework is outlined here.

A. Provide Full Disclosure of Data Collection
  A.1 Description Requirement
  A.2 Enforceability Requirement
  A.3 Irrevocability Requirement
  A.4 Intelligibility Requirement
B Require Consent to Data Collection
  B.1 Acknowledgment Requirement
  B.2 Opt-In Requirement
C Minimize Collection of Personal Data
  C.1 Functional Requirement for Collection
  C.2 Distributed Processing Requirement
D Minimize Identification of Data with Individuals
  D.1 Nonattribution Requirement
  D.2 Separate Storage Requirement
E Minimize and Secure Data Retention
  E.1 Functional Requirement for Retention
  E.2 Security Requirement
  E.3 Nonreusability Requirement

9.6.1 Provide Full Disclosure of Data Collection

Disclosure has long been recognized as a critical element of public data collection, and was a prominent component of the Fair Information Practices. It has been proposed that the disclosure of data collection in the specific context of an information network should take the form of a public statement that personal data will be collected. The statement should then provide a complete description of the type of data to be collected.

The first element of disclosure is the description requirement. An adequate description includes the type of data to be collected; this should be very specific, including details such as resolution or granularity. An adequate disclosure of resolution would specify the granularity, for instance, how often power consumption data samples are being taken.

There must also be a clear indication as to how long and the means by which the data are retained. A focus on the means for retention opens the possibility for the advancement of privacy-aware storage technologies (for example, technologies that limit or prevent data reuse, or technologies that allow a subscriber to retain control over his or her data). Informed subscribers may prefer one storage technology to another, motivating operating companies through market forces to adopt the preferred technologies. Finally, the description should also include the use to which the collected data will be put.

The effectiveness of a disclosure requirement is strongly dependent on an enforcement requirement. The threat of punishment must be of sufficient magnitude and certainty that a collecting entity is motivated to provide a clear disclosure and to comply with it.

It is also important that a given data set is always to be treated according to the privacy policy under which it was collected. An irrevocability requirement should be applied to collected data so that the customer has some certainty as to how collected data are treated for the duration of retention.

The extent to which a subscriber feels secure in his or her communications depends in part on that subscriber's understanding of the data collection disclosure; if the disclosure is too long or written in arcane legal language, the subscriber will be left unsure as to how personal data will be treated. It follows that there must be an intelligibility requirement for data collection disclosures.

9.6.2 Require Consent to Data Collection

The term consent is loaded with legal implications. For the purposes of privacy protection, consent is the flip side of disclosure; it establishes the disclosure as a contract. A requirement for consent also serves a pedagogical purpose: it alerts the user to the presence of data collection and may heighten the awareness of the presence of a potential privacy issue.

It has been proposed that subscribers or users of a given communication technology must acknowledge the data collection disclosure before they can use the technology. The acknowledgment requirement can take the same form as license agreements for software updates. The users must click an appropriate button on a screen before proceeding to use the technology. Such acknowledgments are also found in car GPS units.

The technology that underlies a given service may change over time. A residential consumer may be associated with a given power utility for a long period of time, during which power consumption monitoring technology has changed dramatically. If data collection practices change, the user should be notified. Furthermore, user consent to such alterations should take the form of an opt-in requirement, as opposed to one of opting out. The former clearly increases the extent to which the consumer understands and acknowledges data collection [18].

9.6.3 Minimize the Collection of Personal Data

Personal data consist of information that identifies or is correlated with the behavior, thoughts, and/or preferences of an individual. The first requirement under this heading is probably the most important of all of the design requirements—it is that such collection be necessary. Specifically, there must be a functional requirement for collection: the collection of personal data must be tied to the functionality of the technology.

The distributed processing requirement calls for data to be used as close as possible to the point at which it is collected. There are two rationales for this requirement. First, it prevents the creation of a single database that could be a target for hackers, law enforcement, or others who wish to exploit the data. Second, it reduces the ability of the service provider to market the data to third parties or to reuse the data for purposes other than those for which it was originally collected.

Finally, any data that is collected in bulk for later testing purposes should be aggregated and/or anonymized.

9.6.4 Minimize the Identification of Data with Individuals

There is a distinction to be drawn between collecting information about equipment and collecting information about the equipment's users. The first subrequirement under this heading is the nonattribution requirement, which calls for anonymizing the data collected about equipment wherever possible. Given that many services are billed to individuals, there must be some connection between the usage of the service and personally identifying information, such as a name and address. We propose a separate storage requirement such that billing and functional records are stored in separate places.

9.6.5 Minimize and Secure Data Retention

Data retention must be directly related to the functionality of the technology. It is not sufficient that data is useful in some other context or may be useful at some future date. A functional requirement for retention has been proposed: the storage of the data must be directly connected to the functionality of the technology. If data must be stored, then it must be protected. The basic security requirement requires that data be stored in such a manner that inadvertent disclosure is difficult or impossible. This is a long-standing, general concern in many industries, so we do not dwell on it here, except to note that a requirement that consumers be notified when data is lost or stolen has been shown to reduce the frequency of such events.

Finally, but perhaps most importantly, there is a nonreusability requirement: it calls for data to be stored in such a manner that its use in an undisclosed manner should be difficult or impossible.

9.7 Privacy Problems and Advanced Sensing System: Demand Response Case Study

Demand response provides real-time electricity pricing information to consumers to motivate consumption behavior that results in a more uniform demand distribution. The basic idea is that by showing consumers the cost of their consumption behavior, they are more likely to perform electricity-intensive domestic tasks during off-peak hours. As a consequence, demand levels out, alleviating the utilities' (expensive) need to take generators online and off-line over the course of the day.

Demand response systems require power consumption information at a level of granularity far finer than that required for monthly billing. The solution lies in advanced metering infrastructure (AMI), a technology that can sample and record power consumption on a minute-by-minute basis, as opposed to the once-a-month meter readings of the past.

9.7.1 Power Consumption Data Creates a Privacy Issue

TRUST researchers at Cornell University have shown that the detailed power consumption data collected by advanced metering systems reveals information about in-home activities. Furthermore, such data can be combined with other readily available information to discover even more about an occupant's activities [19]. In the latter reference, an experiment conducted in a standard student residence is described.

An energy usage monitor manufactured by Brultech, Inc., was attached to the residence's breaker panel to collect real-time power consumption data. The data, obtained at intervals of 1 or 15 second(s) with a resolution of 1 watt, was transferred to a nonintrusive load monitor (NILM) application running on a workstation. A behavior extraction algorithm was then run on the workstation in an attempt to predict behavior based solely on power consumption. Video data was used to establish a control for the experiment.

Some of the results from the experiment are reproduced in Figure 9.8. Figure 9.8(a) depicts aggregate power consumption data over the course of several days. The vertical axis is labeled in watts, while the horizontal axis depicts the passage of time over the course of several days. There are several substantial power consumption peaks over the course of each day, indicating activity within the residence.

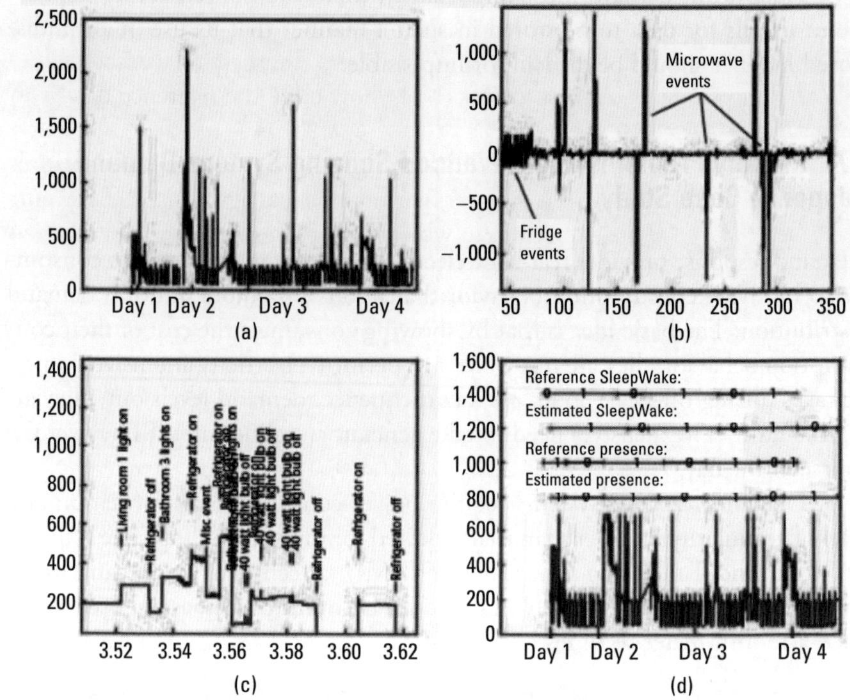

Figure 9.8 Behavior-extraction algorithm. (a) The aggregate power-consumption data; (b) the derived switch events; (c) a screenshot of the load identifier; and (d) a comparison between reference and estimated intervals [19].

Figure 9.8(b) illustrates the results of an edge detection algorithm applied to the power consumption data collected over several hundred seconds. The edge detection algorithm is quite simple—the graph depicts the difference between power consumption samples that are adjacent in time. The vertical axis depicts $\Delta(t) = P(t) - P(t-1)$, where $P(t)$ is the power consumption sampled at time t. The horizontal axis reflects time. Note that certain switching events can now be isolated: the power consumption transients created by a refrigerator and a microwave oven are easily seen.

Figure 9.8(c) is a screenshot taken from a load-identifying program. It shows how events can be isolated and classified over the course of a day (the units of the horizontal axis is days). With this type of information, we can proceed to estimate the behavior of the individuals within the home.

Figure 9.8(d) shows that power consumption data can be used to estimate variables related to personal behavior. The vertical axis is labeled in watts, while the horizontal depicts the passage of time over the course of several days. The reference lines show actual behavior. On the "Reference SleepWake" line, 0 indicates that the occupant was asleep, and 1 indicates that he was awake. On the "Reference presence" line, 0 indicates that the occupant of the residence was not at home, and 1 indicates that he was at home. The estimated lines indicate our estimates of these events. Note how close the reference data is to the estimates.

Table 9.2 shows the results of running the behavior estimation algorithm on a 3-day training set and a 7-day experimental set. Several specific events were selected to indicate the extent to which power consumption data revealed behavior in the home. These include the presence of one or more individuals ("presence"), whether or not the residents were asleep ("sleep cycle"), the operation of the microwave oven ("microwave"), and the operation of lights within the residence ("bathroom lights," "passage lights," and "living room lights"). For each quantity, Table 9.2 shows the number of events recorded, the percentage of successfully detected reference events, the misdetection percentage, and the percentage of the reference interval correctly classified.

9.7.2 A Privacy-Aware Demand Response System

When demand response systems are considered from the perspective of privacy-aware design, a privacy-preserving solution is apparent. The desired behavior—

Table 9.2 Performance of Behavior Extraction Algorithm

	Sample Size	Reference Events Detected	Percent Misdetection	Percent Interval Correctly Classified
Training Data				
Presence	8 Ref., 8 Est.	100%	0%	97.3%
Sleep cycle	6 Ref., 6 Est.	100%	0%	93.4%
Microwave	8 Ref., 8 Est.	50%	78%	43%
Bathroom lights	8 Ref., 8 Est.	72%	44%	52%
Passage light	8 Ref., 82 Est.	38%	90%	57%
Living room lights	8 Ref., 8 Est.	55%	88%	58%
Experimental Data				
Presence	10 Ref., 10 Est.	80%	20%	97.4%
Sleep cycle	12 Ref., 10 Est.	83%	0%	92.3%
Microwave	10 Ref., 58 Est.	80%	83%	99%
Bathroom lights	60 Ref., 103 Est.	63%	42%	81%
Passage light	8 Ref., 82 Est.	38%	90%	57%
Living room lights	19 Ref., 179 Est.	21%	89%	52%

Source: [19].

cost-aware power consumption by consumers—is highly distributed. Demand-response is predicated on the presentation of pricing data in homes across the country. With the distributed processing requirement in mind, it becomes clear that it is not the power consumption data that needs to be collected, but it is instead the pricing data that needs to be distributed. Fine-grained consumption information never needs to leave the home, thus alleviating any need for privacy concerns.

Demand response systems call for several data flows. For each of them, a privacy-aware approach can be adopted. First, pricing data must be presented to the consumer so that he or she has a basis upon which to make consumption decisions. This does not present a privacy concern, as the utility can simply broadcast the pricing to the residential meter and/or to an application on the consumer's home computer.

The second flow is more problematic. Consumer-specific consumption data must be provided to the utility for billing purposes. There is an issue here, as one cannot stream consumption data to the utility without creating the aforementioned privacy issue. One also cannot stream real-time cost data, as it would be trivial to convert this information back into consumption data. The solution lies in accumulating price-weighted consumption data at the residence and then sending the aggregate cost to the utility on a weekly or monthly basis. This implies a level of security at the meter that requires a trusted platform module or the equivalent.

Finally, the utility needs consumption data, temporally precise, but aggregated at the level of the consumer, in order to predict demand and maintain a price model. Typically aggregated real power consumption data at the substation level is sufficient to predict the need for new transmission and distribution lines and generation necessary to service the predicted demand. A neighborhood aggregator can be used to combine and anonymize data so that the desired temporal granularity is provided without generating information about individual behavior. Aggregator contractual obligations to the utility provides it with information sufficient to determine how much of the predicted demand can be mitigated through pricing mechanisms. In any case the utility's need for consumption data should not be at the level of the individual consumer. Anonymization can be performed by summing the power consumption data for a sufficient number of customers so that a single customer's data cannot be isolated.

These approaches result in the architecture depicted in Figure 9.9, which shows power consumption data being collected from the residences on the right. Though shown as separate blocks, the AMIs are attached to the residences in such a manner that the raw consumption data cannot be accessed without entering or being very close to the home. The AMI transmits encrypted fine-granularity consumption data to the neighborhood aggregator, which decrypts

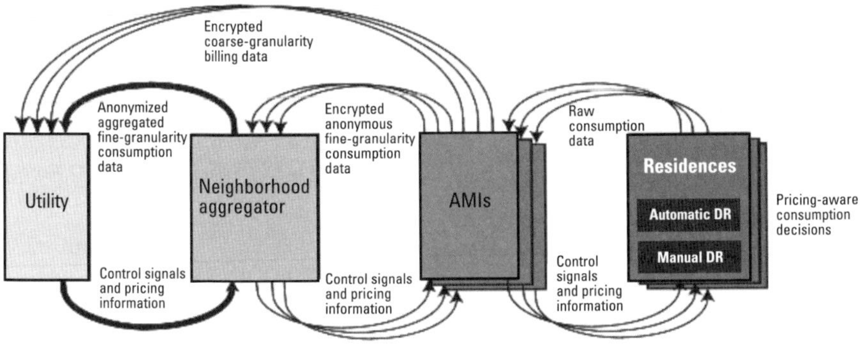

Figure 9.9 Privacy-aware demand response architecture [20].

the data and combines it with data from other residences. The resulting anonymized data is passed to the utility on the left. The rightward flows (shown at the bottom of the figure) depict control and pricing information sent from the utility to the aggregator, AMIs, and residences.

9.8 Conclusion

Sensor networks offer a powerful means for securing and monitoring critical infrastructure. However, as shown for the case of power consumption data, data collection often creates a privacy problem for the individual, and a public relations problem for the collecting authority. It should be noted that there are many other cases in which collected data—important to the functionality and protection of critical infrastructure—creates similar privacy concerns. The privacy-aware design rules discussed in this chapter provide an important framework for deploying advanced sensing systems to protect critical infrastructure. With the appropriate concern for privacy and creative design, it would be possible to exploit novel sensing technologies for the public good.

References

[1] Goldsmith, A., and S. B. Wicker, "Design Challenges for Energy-Constrained Ad Hoc Wireless Networks," *IEEE Wireless Communications Magazine,* Vol. 9, No. 4, 2002, pp. 8–27.

[2] Zhang, X., and S. B. Wicker, "On the Optimal Distribution of Sensors in a Random Field," *ACM Transactions on Sensor Networks,* Vol. 1, No. 2, 2006, pp. 301–306.

[3] Burnham, M. R., et al., "Biological Functionalization and Surface Micropatterning of Polyacrylamidehydrogels," *Biomaterials,* Vol. 27, No. 35, 2006, pp. 5883–5891.

[4] Chandrashekhar, M. V. S., et al., "Demonstration of a 4H SiCbetavoltaic Cell," *Applied Physics Letters*, Vol. 88, No. 3, paper 033506, 2006, pp. 1–3.

[5] Ekanayake, V., C. Kelly IV, and R. Manohar. "An Ultra-Low-Power Processor for Sensor Networks," *Proceedings of the 11th International Conference on Architectural Support for Programming Languages and Operating Systems*, Boston, MA, October 2004.

[6] Giani, A., et al., "A Testbed for Secure and Robust SCADA Systems," *SIGBED Review*, Vol. 5, No. 2, July 2008, pp. 1–4.

[7] Downs, J. J., and E. F. Vogel, "A Plant-Wide Industrial Process Control Problem," *Computers & Chemical Engineering*, Vol. 17, No. 3, March 1993, pp. 245–255.

[8] Warren, S. D., and L. D. Brandeis, "The Right to Privacy," *Harvard Law Review*, Vol. 4, No. 5, 1980, pp. 193–220.

[9] Cooley, T. M., *A Treatise on the Law of Torts*, 2nd ed., Chicago, IL: Callaghan, 1888.

[10] Goffman, E., *The Presentation of Self in Everyday Life*, New York: Doubleday, 1959

[11] Bentham, J., "The Panopticon; or, The Inspection House (1787)", in *The Panopticon Writings*, M. Bozovic, (ed.), London, U.K.: Verso, 1995.

[12] Foucault, M., *Discipline and Punish*, New York: Vintage, 1995, English translation (*Surveiller et punir: Naissance de la Prison*, Paris: Gallimard, 1975).

[13] Cohen, J., "Examined Lives: Informational Privacy and the Subject as Object," *Stanford Law Review*, Vol. 53, 2000, pp. 1373–1437.

[14] Gandy, O., *The Panoptic Sort: A Political Economy of Personal Information*, Boulder, CO: Westview Press, 1993.

[15] Solove, D., *The Digital Person: Technology and Privacy in the Information Age*, New York: NYU Press, 2006.

[16] Wicker, S. B., and D. E. Schrader, "Privacy-Aware Design Principles for Information Networks," *IEEE Proceedings*, Vol. 99, No. 2, February 2001, pp. 330–350.

[17] *Records, Computers, and the Rights of Citizen*, HEW, 1973.

[18] Sovern, J., "Opting In, Opting Out, or No Options at All: The Fight for Control of Personal Information," *Washington Law Review*, Vol. 74, 1999, pp. 1033–1130.

[19] Lisovich, M., D. Mulligan, and S. B. Wicker, "Inferring Personal Information from Demand-Response Systems," *IEEE Security and Privacy Magazine*, Vol. 8, No. 1, January/February 2010, pp. 11–20.

[20] Wicker, S. B., and R. J. Thomas, "A Privacy-Aware Architecture for Demand Response Systems" *Proc. of the 44th Hawaiian Conference on System Science (HICSS-44)*, Kauai, HI, January 10, 2011, p. 130.

10

Military Defense, Civil Defense, and Civil Protection Integration in a Multiscenario Crisis Event

Roberto Mugavero

10.1 Introduction

The experience carried out during the planning organization and management of the Major Event G8 Summit 2009 L'Aquila has some special and unique characteristics at an international level. Until now, we have always been accustomed to deal with the methodologies and systems complexity regarding safety and security issues (management of natural disasters, like earthquakes, or international events such as a G8 summit), but generally in separate areas and contexts both spatially and temporally. In the case of the L'Aquila events, however, we found ourselves in the position to share, in terms of both space and time, the need to manage both the emergency related to an earthquake of a significant size, magnitude, and effect and the security framework linked to an international summit that was attended by a large number of personalities. This has, for the first time, required the collation, integration, and sharing of action planning, organization, and management for human and instrumental resources with different aims and objectives all housed in the same local context and where the effects and the consequences of an event were directly affected from the contiguous event.

This experience has led, for the first time, the development of a methodology and a model of integration in which civilian and military capabilities, aimed at two different aspects of safety and security, have joined forces to create a unique and homogeneous framework in which two events took place in a parallel but simultaneously converging manner. For this reason we can say that the Italian experience is a case study of exceptional importance both for instrumental and methodological solutions adopted and for the significance of the events that took place.

This chapter is focused on the study of presenting security framework of the Italian G8 Summit 2009, the adopted methodologies and technologies, the particular difficulties, and the implemented solutions, related to the summit displacement from the city of La Maddalena to the city of L'Aquila in a complex scenario: little time available to rearrange the new devastated summit site, the characteristics of locations, the geographical conditions, and the presence of a territory by the effects and consequences of an earthquake.

The study of the summit security planning and organization is also important considering the fact that the event was placed in a delicate political moment characterized by the presence of a serious international conflict and a severe economic and financial crisis. The scenario has led to the consideration of, in addition to the traditional forms of turmoil against the G8 Summit implemented by national and European movements, novel forms of dissent related to global politic-economic situations. In addition, the fact must be added that the summit would represent a potential opportunity to focus the attention on the region (Abruzzo) situation after the earthquake.

10.1.1 Major Events

The definition of a major event includes those initiatives, single or repeated, that require significant investment, considerable planning organizational efforts, the adoption of marketing techniques, and the participation of visitors, public, and media.

The theme of this type of initiative may be sporting events (Summer and Winter Olympic Games and the World Cup), commercial events (auto and nautical shows), cultural events (art exhibitions, film festivals, music festivals), and religious (Jubilee) or political/economical events (international summits, G7/G8, WTO).

In general, the main feature of the major events is their unique and exceptional profile, in addition to the fact that they operate for a limited time not comparable to the time required for their organization and preparation. Thus, those events are characterized by having a major impact, both in their preparatory phase and during their progress. This makes particularly complex and sensitive the organizational phase in terms of public order, mobility, health, accommodation, and, most importantly, security.

In fact, as we know, at international level the general terrorist threat framework changed with the fall of the Berlin Wall and the World Trade Center attack. Since the last century the dimension of terrorism risk is increased due to the combination of additional availability of technology, expertise, and means. This asymmetric threat generates a lot of areas of instability, due to local tensions, social, ethnic, and religious conflicts, and the presence of terrorist networks and organized crime.

Italy, as well as many other western countries, is increasingly exposed to potential risk of terrorist attacks because of changing of international geopolitical framework in recent years, characterized by instabilities. This risk is even greater for the Italian peninsula, because it is the weapon to strike Western symbols and values, of which Italy is highly representative due to its historical and artistic heritage and the presence of the Pope. Thus, major events are increasingly exposed to potential risk of terrorist attacks.

For these reasons it is crucial to obtain the best result in the event security management through coordination and synergy of entities, institutions, individuals, organizations, resources, skills, knowledge, training, and the means involved in all the security activities.

10.1.2 Prospects and Problems

While the major event has a limited lifetime, the implications associated with it can have long-term effects during its preparation phase, also generating permanent consequences physically linked to the territory. This leads to the introduction of the concept of inheritance of an event, an issue that must be properly designed and planned: symbols, values, messages, and the operations of direct and indirect marketing planning.

The inheritance of a major event cannot be uniquely defined and post-event effects are not readily determinable. Moreover, these effects are not always positive, and it is difficult to foresee reliable data and information on possible negative effects.

In any case, the goal of the host city is to get a positive image. This element is very important for the local authorities and the national governments, but also for the international organizations that plan and/or support the event, mainly for three reasons. A positive inheritance is proof that the event was good for the host city/nation, is a demonstration of the proper use of funds for infrastructure, and is the best motivation for the candidacy of the local territories/nation for other events.

Based on lessons observed, it is possible to identify some key themes that characterize an event of this scale and therefore may represent the lowest common denominator to try to anticipate/predict the effects of a major event: security, urban regeneration, infrastructure, services and logistics, territorial economic development, investment and marketing, international political legitimacy, skills and education development, and environmental impact.

10.2 Italian National Protection System

To understand how the Italian major events security is planned, organized, and managed, it is fundamental to know how the organization of the Italian national territorial protection system is planned [1].

The main activities of the national protection system are:

- Mitigation or risk assessment;
- Preparedness or planning scenarios and security tasks;
- Response or national emergency response system;
- Recovery or consequence management.

10.2.1 Risk Assessment

The risk assessment activity integrates the findings of the intelligence and law enforcement with input from the scientific and technical communities and reflects suspected terrorist capabilities and known threats; the result of key findings in this assessment can be shown in a table, linking probability and consequences, and should be updated in time and when appropriate.

10.2.2 Planning Scenarios

The main Italian threat scenario can be categorized as:

- The national plan against terrorist attack;
- The response plans set up by the single state agencies;
- The provincial civil defense emergency plans.

10.2.3 Security Tasks

Security tasks are divided between security tasks in Italy and security tasks abroad.

Security tasks in Italy are mainly classified among infrastructure and sites protection, borders and coastlines surveillance, road and transport protection, networks and public protection, nonproliferation, and military and civilian operations. Security tasks abroad (Figure 10.1) are mainly divided into disarmament and nonproliferation and military and civilian operations.

The general tasks, common for internal and abroad activities, are to ensure C3I: Command–Control–Communications and Information.

10.2.4 National Emergency Management System

The national emergency management system is composed of two structures: civil defense and civil protection.

The civil defense task is the safeguard of human life, goods, national heritage, human settlements, and the environment from the consequences of war or terrorist attacks.

The civil protection task is the safeguard of human life, goods, national heritage, human settlements, and the environment from the consequences of each kind of natural or man-made nonintentional disaster.

Therefore, the Italian homeland protection is a system where there is an activation of one of the two structures depending on the type of danger:

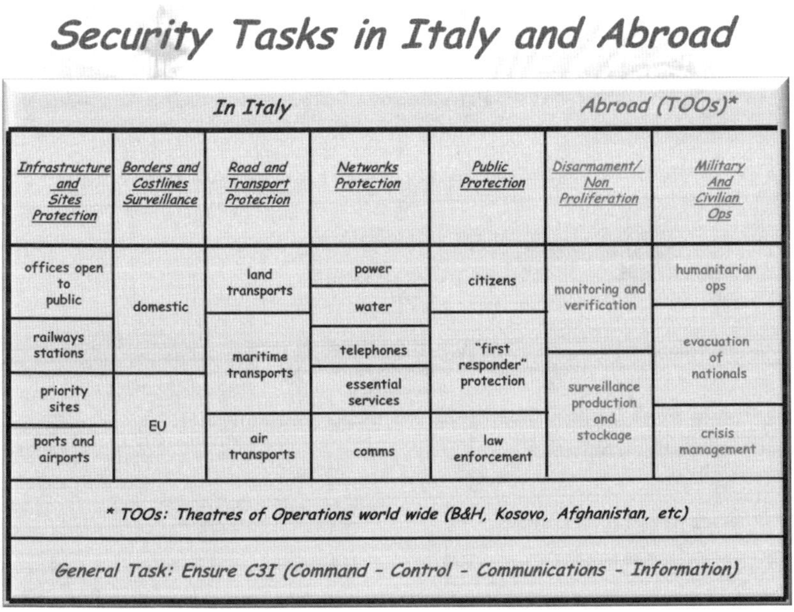

Figure 10.1 Security tasks.

- For a military or terrorist event, the civil defense system is activated.
- For natural disasters or accidents, the civil protection system is activated.

The Italian civil defense organization involves all the actions to protect the civilian population and public/private facilities in the following emergencies: national/international crises and conflicts.

The civil defense is composed of six areas of intervention: continuity of government action, public information and training, telecommunications alarm systems, public health, and economic system protection.

As shown in Figure 10.2, the national civil defense organization is structured as a chain, where command and control and national decisions are taken by the Presidency of the Council of Ministers supported by decision and coordination committees.

The operating national coordination is guaranteed by the Ministry of Interior on the basis of two structures: the Crisis Unit, chaired by the head of police, for security management, and the Inter-ministerial Technical Committee for Civil Defense, chaired by the head of the national fire brigade, for safety management. On-site assistance coordination is guaranteed by the local

Figure 10.2 Italian civil defense organization.

prefect and key operations are guaranteed by the police forces, armed forces, fire brigade, and emergency health services.

In summary, the Italian civil defense system involves in its actions several national public organizations, developing preventive activities and coordination during the emergency: the presidency of the Council of Ministers, the Civil Protection Department, the Ministry of the Interiors, the prefect, the head of police, the police forces, the armed forces, the fire brigade, the National Health Service, the Italian Red Cross, hospitals, the municipal authority, public health and environmental protection services, and essential services.

The civil protection organization is based on four different levels of emergency: national, regional, district, and municipal. They have specific administrative areas that are connected with certain types of events, as shown in Figure 10.3. These events can be categorized as follows:

- Events that can be solved by applying the available resources from the local community;
- Events that need additional coordinated resources from outside the local community;
- Events that, due to their intensity and size, should require extraordinary means and authorities.

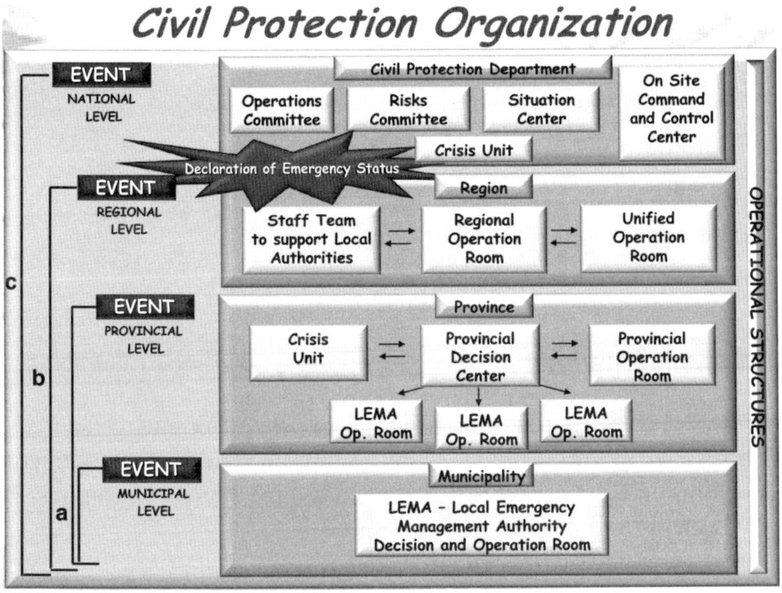

Figure 10.3 Italian civil protection organization.

The organization is composed by different levels of competence corresponding to the particular kind for the extension and magnitude of the event. However, it is interesting to note the presence of the rule of subsidiarity, according to which, in case of a lack of action by a level of competence, the next higher level is automatically activated.

The Italian national system of civil protection was established in 1992 and involves many different entities: the National Civil Protection Department, which includes the Presidency of the Council of Ministers and has coordination functions, the fire brigade, the police forces, the armed forces, the National Health Service, the Italian Red Cross, hospitals, the municipal authority, public health and environmental protection services, civil protection volunteers, Alpine and Speleological Rescue, regions, provinces, municipalities, and mountain communities, national scientific research groups, the Geophysics and Volcanology Institute, national technical services, and essential services.

10.2.5 Recovery and Consequence Management

The consequence management activity includes all the operations that must be taken as a result of an event in order to mitigate the effects to humans and the environment. Consequence management plays a vital role in safeguarding the health of the community and includes all the needed interventions, made by public and private organizations, for removing the obstacles to the rehabilitation and restoration of normal life conditions.

The main involved entities and structures are:

- National authorities;
- Regions;
- Provinces;
- Municipalities.

10.3 System Criticality

The analysis of the activities during the Italian past major emergencies and disasters has shown in some cases a lack in the appropriate event management and problem solution. The reason is that there are some system criticalities within both civil defense and civil protection responses.

There are different types of criticality:

- The first criticality originates by the ever-increasing worldwide number of terrorist attacks, also involving chemical, biological, radiological, and nuclear (CBRN) or dual-use agents, for number, magnitude, and damage effects.

- At the same time, in the second criticality, there is a general reduction of human resources available for operational countermeasures. In Italy this reduction originates from the abolition of compulsory military service in 2004, determining a severe decrease of personnel to deploy on the ground during a crisis event.
- The third criticality originates because the Italian territorial security management system is composed by a lot of different entities, resources, and tools that should interact, and it is not easy to perform actions of mitigation, preparedness, response, and recovery.

These criticalities determine effects on all national emergency response systems, and the main consequences are generated on strategic planning and operational activities. In order to balance the negative consequences of system criticality, corrective actions must be implemented.

First, organization and management must be optimized. This can be achieved mainly through a heterogeneous resources sharing, an integrated security policy strategy, an increase of risk knowledge, a definition of a clear and exhaustive risk framework, and an optimization of planning actions.

Then we have to develop new operational capabilities to obtain better event management. These capabilities should be dynamic, fast, effective, scaled-down respondents in case of multiple scenario events and they should have the integration of military and civil capabilities. All these actions are intended to achieve an integrated response system where national defense, composed by military defense and civil defense/civil protection, works closely with a high level of coordination involving various entities and operational structures.

The technological criticality compensation needs an evolution of organization, technology, and system architecture. As a matter of fact, today this architecture is based on different layers that do not allow the enhancement and homogeneity of field organization with process management and technology areas with data acquisition by implementing fusion, processing, and the decision support system in order to forecast with scenario evolution.

In the future a systemic approach is imperative where infrastructures and processes allowing the organization of logical and structural architectures lead to an integrated system. As a matter of fact, by using a service-oriented architecture, we can reproduce, in a single system, data, services, and processes from different sources, thus allowing the creation, integration, and coordination, in real time, of the right responses with a continuous matching with the evolving situation. A systemic approach also concerns the operative response integration.

For the time being, we have a low cooperation system composed by several organizations with separated regulations and administrative actions, low integration, and coordination between organizations, separated by different methodological and technological standards. In the future we need a strong, high-cooperation system using a common methodological and technological standard. This process of integration must be accompanied by a system engineering evolution.

The current approach is that each system is designed according to specified parameters and performances. The evolution must lead to an approach where, starting from the full complex of requirements, considering the operational scenarios, and making use of modeling and simulation, a system of systems approach is implemented that meets the overall performance requirements. The aim of all these actions must be the creation of a homogeneous and collaborative environment allowing an institutional, geographical, technical, and operational connection to achieve a rational, coherent, and interoperable response. The key point is that this is not only a technical problem, essentially manageable, but also an operational problem at an institutional level, once more difficult even to define and then solve. The reason is that the human being is not a computer; making a computer grid is easy, which is not the case for making a human-being grid.

10.4 Summit G8 Security Planning

The aim of this section is to illustrate and describe all the structures, resources, and tools adopted to ensure the G8 Summit Security in L'Aquila, Italy, July 8–10, 2009.

As already stated in Section 10.1.1, a major event is an event characterized by organizational complexity in terms of security, public order, mobility, housing, welcome, and healthcare, with the presence of a large number of personalities and/or participants, with the inevitable need to take extraordinary and urgent measures.

In Italy, this kind of event can be managed with the same tools and resources used in the management of emergencies due to natural or nonintentional man-made disasters. The declaration by which the event is classified as a major event occurs with a decree by the President of the Council of Ministers. The decree normally contains the declaration, the duration, and extension in close relationship to the quality and nature of the event and the appointment of a delegated commissioner to ensure the coordination of activities. Its implementation is enforced by means of orders notwithstanding any provision in force and in respect of the general principles of law.

The G8 Summit was declared a major event, with a Presidency of the Council of Ministers Decree in 2007 [2, 3], under the responsibility of the Department of Civil Protection as other exceptional occurrences. The headquarters of the summit was initially identified on the Island of La Maddalena in Sardinia.

By a Prime Minister's Order, the head of the Civil Protection Department was appointed as the delegated commissioner for the coordination of interventions and initiatives related to the summit organization.To carry out this activity, the delegated commissioner has used the structure of the Mission of the Presidency of the Council of Ministers, consisting of the staff of the Civil Protection Department and the presidency of the Council of Ministers, as well as military and civilian personnel employed by other state departments and public or nonterritorial institutions.

On April 6, 2009, at 3:32 a.m., the city of L'Aquila was hit by an earthquake of a magnitude 5.8 on the Richter scale (see Figure 10.4). The city is located in the Italian region of Abruzzo, and an idea of the intensity of the earthquake can be graphically deduced by looking at the photo where a badly damaged area in the center of Onna (only a few kilometers from L'Aquila) is shown in Figure 10.5.

Figure 10.4 Earthquake map.

Following the earthquake of April 6, 2009, the displacement of the G8 Summit to L'Aquila was decreed [4] in the framework of solidarity to the affected territories and improvement of devoted economic resources for the region recovery. With the decree, the delegated commissioner was responsible for arranging the reprogramming the organization of the summit and for taking any action required by the new location of the international conference. The result was an integrated management of the earthquake emergency and the organization of the G8 Summit, as shown in Figure 10.6.

The integrated management of the Abruzzo earthquake emergency and the G8 summit organization has, as the main point of connection, the head of the Civil Protection Department as the delegated commissioner for both events. For the operational direction of the two events, the Financial Police School of Coppito (L'Aquila) was chosen as both earthquake emergency headquarters and the G8 Summit headquarters. The integrated management determined two lines of activities, parallel but separate, concerning the two events.

10.4.1 Participants

The event has been very complex for the type and number of participants and for the complexity of organization in terms of security, law enforcement, mobility, hotel reception, hospitality, and healthcare. The following quantitative numbers are of interest:

- Attending delegations: 39;

Figure 10.5 Badly damaged area.

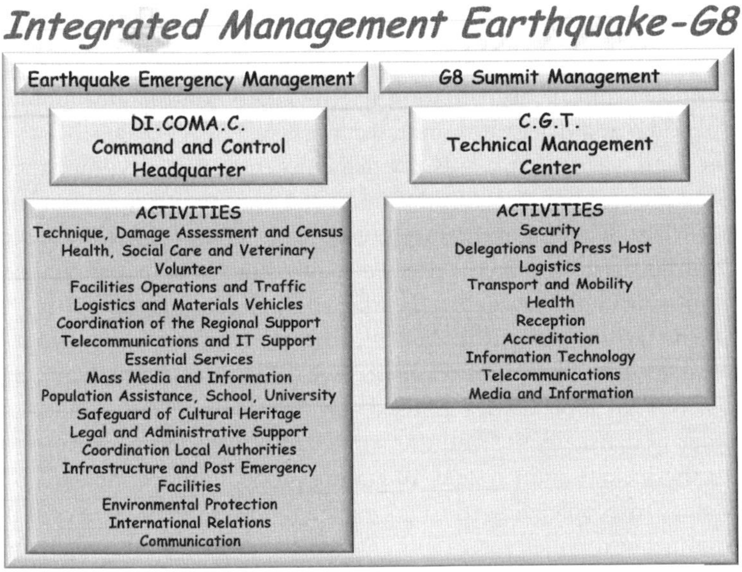

Figure 10.6 Summit organizational structure.

- Participant delegates: 4,500;
- Media representatives: 4,500;
- Security personnel: 15,000;
- Staff personnel: 2,000.

10.4.2 Planning and Organization

For the G8 Summit security system planning and organization, the main targets were to guarantee the protection of all the participants in the summit, normally conduct summit activities, retain civil rights to demonstrate and to express ideas in a pacific way, and harmonize necessities of the summit security with the livable safeguard of the L'Aquila district with the consequences following the earthquake. The operative planning was built from an ordinance of the Prime Minister and a decree of the Ministry of Interiors [5–16]. The final design of the security plan was made by the county committee for order and public security.

The main data accounted for the planning business were: delegation numbers and composition, territory characteristics and classification of involved areas, viability, place conditions following earthquake consequences, summit agenda, summit venue features, accommodation features, delegation itinerary, participant transfer system, and press transfer system. These data were used to plan the summit land and air security.

10.4.2.1 Land Security

Land security planning was divided mainly into the sections of land security areas, territorial control, accreditation system, access control, area mobility, exchange parking, and delegation route security.

The land security area was divided into three zones: compliance area, comprising the area not directly affected by the summit activities, but involved in the participants land movements; reserved area, comprising the areas immediately behind the zone directly affected by the summit activities; and security area, comprising all the areas directly affected by the summit activities.

For the territorial control [17–19], to prevent criminal acts and to guarantee event security, the resources used were police and Army, Navy, and Air Force, whereas the tools used were fixed and mobile surveillance, video surveillance, radar systems, aircraft, and unmanned aerial vehicles.

For the accreditation system, to grant a proper authorization to go inside security areas to people allowed to go in, the used resources were civil protection personnel and police forces units, whereas the used tools were computer systems.

For access control, to prevent unauthorized people and vehicles a presence within the areas, the used resources were police and armed forces, while the used tools were mobile and fixed surveillance, metal detectors, and radiogenic equipment.

For area mobility, to guarantee the area mobility of delegates, residents, information staff, authorized service staff, and, in case of emergency, rescue vehicles, the used resources were police and armed forces, whereas the used tools were mobile and fixed surveillance, metal detectors, and radiogenic equipment, similar to the access control rules.

For exchange parking, to guarantee to delegates, residents, information staff, and authorized service staff vehicles parking within the traffic interdicted areas, and to guarantee the transport from/to traffic interdicted areas, the used resources were police and Army, Navy and Air Force, as well as private security, whereas the used tools were mobile and fixed surveillance, metal detectors, and radiogenic equipment.

For delegations routes security, to guarantee the security of the delegations' members, the used resources were police and armed forces, whereas the used tools were protection services and mobile and fixed surveillance along the routes.

10.4.2.2 Air Security

Regarding air security planning organization, this was mainly divided into the air security area and the delegations' air transfer and event security.

For the air security areas, the involved zone was divided in two areas: zone 2 with interdiction of all the visual flight air traffic and zone 1 with interdiction of all the air traffic.

For the transfer of delegations and event security, to guarantee protection from criminal actions coming from the air, the resources used were police and Army, Navy and Air Force, whereas the used tools were mobile and fixed surveillance, aerial means, radar systems, and anti-aircraft systems [20].

10.4.2.3 Special Security Activities

Due to the presence of many participants and personalities, the complexity of the scenario, the characteristics of locations, the geographical conditions, and the presence on the territory of the effects and consequences of the earthquake, particular importance has been also devoted to the adoption of special security activity and measures.

As stated in Section 10.1, the event took place during a delicate political moment, characterized by the presence of serious international conflict and a severe economic and financial crisis. This scenario has led to the consideration of the traditional forms of the turmoil by national and European movements and even of forms of dissent relating to global politic-economic situation. It must be also added that the summit would represent a potential opportunity to focus the attention on the situation of Abruzzo after the earthquake.

The special security activity and measures were essentially devoted to prevent illegal actions and terrorist attacks; intervene in case of crisis caused by terrorist actions, even with use of chemical, biological, radiological, and nuclear (CBRN) agents; guarantee surveillance, control and security into traffic mobility from/to areas of event; guarantee, in relation to the event, the national territory incoming control; prevent the execution of illegal acts through mobile phone systems, public and private broadcasting, institutional telecommunications networks, amateur radio, and Internet networks; guarantee security of the delegations' members and the territory surveillance and control; and acquire documents to legal purposes. For all these activities, the used resources and tools were police and Army, Navy and Air Force, special units, firefighters and health services units, aerial means, unmanned aerial vehicles, mobile and fixed video surveillance systems, and CBRN means and equipments.

10.4.2.4 Use of Armed Forces in Security Activities

Within the activities of supervision, control, and security of the areas affected by the summit, a consistent use of manpower, resources, and systems of the Army, Navy, and Air Force was provided. This choice was dictated by the organizational level, the instrumental and human resource equipment, and the available levels of training and expertise.

By the order of the President of the Council of and the Chief of Police Directive, regulated coordination between the police and the Army, Navy, and Air Force has also been implemented including the levels of relative responsibility. For this purpose, military personnel took part in a special course, organized by the General Command of Carabinieri (the Italian Army Police), to qualify them as law enforcement officers.

The military activity was mainly divided into the terrestrial defense component and the air defense component.

The aim of the terrestrial defense was to guarantee territory surveillance and security service. The used resources were army units, whereas the used tools were mobile and fixed surveillance.

The aim of the air defense component was to guarantee the event air space control and defense. The used resources were air forces and army units, whereas the used tools were interceptors and drone, radar, and anti-aircraft systems.

10.4.2.5 Contingency Plans

As part of the activities of planning and organization of the summit security, particular importance has been devoted to the adoption of specific contingency plans: health, natural disaster and/or terrorist attack emergency plans, CBRN, and fire emergency plans were designed. These contingency plans were connected to the general security plan to manage specific emergency situations such as natural (earthquake, fire, and others) and man-made (terrorist attack) ones.

10.4.3 Command and Control Security System

The command and control system was composed by a coordination table, two operation rooms, and two coordination rooms: joint operation room, technical management center, armed forces operation room, and international operation room, as shown in Figure 10.7.

10.4.4 Coordination Table

The coordination table is the structure designed to coordinate the security activity for immediate issues resolution. It is chaired by a delegated commissioner and is composed of several entities.

10.4.5 Joint Operation Room

The joint operation room is the structure for coordination and management of all the resources used for surveillance and public security with operative connection between the territorial and national levels, as shown in Figure 10.7. It was

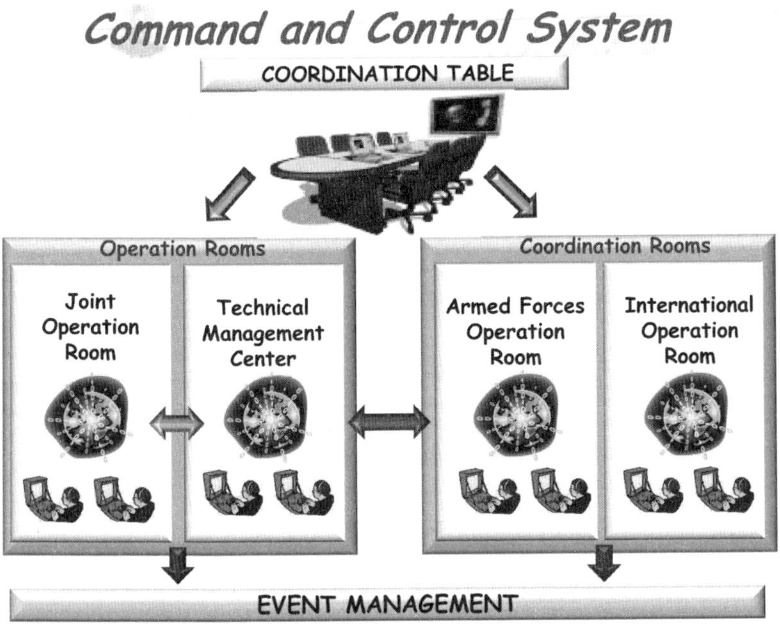

Figure 10.7 Command and control system.

connected with the police main remote headquarters, the main airports, and with the Civil Protection and Ministry of Interiors situation rooms.

The joint operation room, to ensure communication with all remote sites, was connected through voice over IP, intranet, and voice links. Moreover, it was connected, with operational components, through radio networks of each single institution. To ensure the supervision and control of the event, it was equipped, in its position and in remote locations, with software for collection, storage, and management of data, images, and information to share with all the operators.

10.4.6 Technical Management Center

The technical management center, in charge of several functions, was the active instrument of the coordination table decisions for management and coordination of the activities, collection, and sharing of information and coordination and the connection between the organizations activities and surveillance/public security activities.

The technical management center was provided with a decision room and furnished with software for resources/infrastructure display and event management. The center was also connected through voice over IP, intranet, voice links to ensure communication with all remote sites, and, in emergency, through a voice and data satellite backup. Moreover, it was connected, with operational components, through institutional radio, GSM, TETRA, digital mobile radio networks, and optical fibers. The technical management center, dedicated to ensure event supervision and control, was equipped in the central location with software for 2-D/3-D geographical information system for display of resources and infrastructures and for event management.

10.4.7 The Armed Forces Operation Room

The armed forces operation room is the structure for activity coordination of Army, Navy, and Air Force units. This was made through the operation room of the Army, Navy, and Air Force headquarters to coordinate and connect the land command and the air command components.

10.4.8 The International Operation Room

The international operation room is the structure for joint operative and information exchange between national police forces and foreign nation participants to the event. It was composed of an international cooperation team, liaison officers of the summit attending countries, Europol, Frontex, and Interpol.

10.5 Conclusions

The analysis of the activities, during the past Italian major events, has shown in some cases a lack of standard security management and problem solutions that may have occurred during the period of its operation. The analysis of these events pointed out the following requirements as key issues to overcome the problems due to lack of an integrated model of planning, design, and management:

- To share resources, experiences, and methods for all local and national authorities involved in the event management;
- To need, for the operational structures, to perform simple and, at the same time, flexible actions to ensure efficient and enforceable operations upon the occurrence of their needs;
- To acquire accurate and reliable information in order to improve the response times;
- To establish a unique and multidisciplinary approach that can efficiently achieve and develop the proper conduct of the pre-event activities.

With the risk of annoying the reader, a lot of details have been quoted in the chapter. The reason is that all those details point out the complexity of the event and may be useful for assessing some expertise for future similar cases. In the light of above considerations, the Italian G8 Summit held in L'Aquila in July 2009 was a significant effort toward achieving an integrated response security system where military defense, civil defense, and civil protection worked closely with a high level of coordination. In fact, it was a huge field exercise in order to ensure G8 delegations' and participants' security in a territory affected by the consequences following an earthquake: means, resources, tools, and capabilities belonging to various military and civil administrations were used, all interacting among themselves to perform actions of mitigation, preparedness, and response in case of criminal actions or terrorist attacks.

The G8 Summit activity was also a step towards a new technological security system architecture. Technology areas (data acquisition, fusion, processing, and decision support system), including forecasting the scenario evolution and alternative developments, were merged in order to enhance and homogenize field organization with management processes. For these reasons, the study of the Italian G8 security planning, organization, and management is a valuable opportunity to develop a major event security model, consisting of a homogeneous and collaborative environment where institutional, geographical, and technical connection between different administrations and organizations is implemented, based upon a rational, coherent, and interoperable coordinated response.

References

[1] Presidency of the Council of Ministers, Civil Protection Department, Compendium of the main Laws, Decrees, Acts Concerning the Italian Civil Protection System, 2009.

[2] President of the Council of Ministers Decree, Declaration of the Italian G8 Presidency as Major Event, September 21, 2007.

[3] President of the Council of Ministers, Act No. 3629, November 20, 2007.

[4] President of the Council of Ministers, Act No. 1534, April 28, 2009.

[5] President of the Council of Ministers, Act No. 3642, January 16, 2008.

[6] President of the Council of Ministers, Act No. 3660, March 5, 2008.

[7] President of the Council of Ministers, Act No. 3669, April 17, 2008.

[8] President of the Council of Ministers, Act No. 3684 June 13, 2008.

[9] President of the Council of Ministers, Act No. 3690, July 4, 2008.

[10] President of the Council of Ministers, Act No. 3707, October 3, 2008.

[11] President of the Council of Ministers, Act No. 3698, August 29, 2008.

[12] President of the Council of Ministers, Act No. 3710, October 31, 2008.

[13] President of the Council of Ministers, Act No. 3719, December 3, 2008.

[14] President of the Council of Ministers, Act No. 3738, February 5, 2009.

[15] President of the Council of Ministers, Act No. 3742, February18, 2009.

[16] President of the Council of Ministers Act No. 3746, March 12, 2009.

[17] President of the Council of Ministers, Act No. 3704, August 17, 2008.

[18] President of the Council of Ministers, Act No. 3716, November 19, 2008.

[19] President of the Council of Ministers, Act No. 3756, April 15, 2009.

[20] President of the Council of Ministers, Act No. 3696, August 4, 2008.

11

Repel Boarders!

Patrizia Fabbri and Leonardo Quattrocchi

11.1 Introduction

This chapter synthetically presents the piracy scenario, and finally describes an effective and tested system to counter the threat of piracy at sea. A preliminary presentation of the historical behavior of piracy and counterpiracy is given. In the frame of this preliminary presentation, some pictorial inserts are included, to graphically illustrate the various concepts and descriptions.

Then the modern techniques and procedures to counter the modern piracy are described. At variance of the olden days, the counteraction is not implemented by the ship itself, being based only on defense facilities and weapons available on board. Nowadays the defense is essentially based on intense patrols of the ships' routes, early warning of the threat, and rerouting of the ship together with intervention of patrol forces.

As a valuable example, an existing complete antipiracy system is presented in some detail. The two basic components of the defense system are highlighted: the shore segment and the ship segment. The former is devoted to early warning and implementation of the safety intervention, when the threat has been detected; the latter consists of all the instrumentation mounted over the ship to collect the valuable information to be transmitted to the shore segment. This system has been already successfully experimented, and the experiment's results are also referred. Some final conclusions end the chapter.

11.2 Pirates

The English word pirate is derived from the Latin term pirata, and that from the Greek πειρατησ (peirats), that is, brigand, which derives, in turn, from πειραομαι (peiráomai), that is, attempt, the word corresponding to the active verb πειρα (peîra), that is, to attempt.

In our highly technological world, the term piracy grew to include copyright violations of intellectual property, such as music, films, and computer software. These are major crimes and cause heavy economic damage worldwide. However, a resurgence of an ancient type of piracy (i.e., an act of violence or depredation on the high sea) is nowadays much more dangerous. Piracy at sea is a crime with ancient origins. As long as there have been ships at sea, pirates have sought to steal from them.

Various methods are used by pirates to survey the area and identify possible targets; the simplest is visual observation, as shown in Figure 11.1, but they also use modern technology such as automatic identification system (AIS) receivers, intelligence through analysis of ships' routes and cargoes, and triangulation of VHF radio transmission. When target ships are close to shore, attacks are carried out directly launching the boats from the beaches, as shown in Figure 11.2; otherwise, mother ships are used.

The number of acts of piracy and armed robbery against ships, reported to the International Maritime Organization to have occurred or have been attempted in 2009, are estimated to be over 400—an increase of 25% over the figure for 2008—a serious threat not only to sea commerce and shipping, but also to the lives of seafarers.

By June 2010, over 20 ships had been detained by pirates only in the Eastern Africa region, with hundreds of sailors held captive and risking their lives.

Figure 11.1 Visual observation is the simplest method used by the pirates to identify possible targets.

Figure 11.2 When target ships are close to shore, pirates launch their boats for the attack directly from the beaches.

Seizures of ships by pirates off the coast of East Africa increased sevenfold from 2005 to 2009, according to the International Maritime Organization, although naval operations in the area appear to be paying off, producing a significant drop in the number of attacks in the first quarter of 2010. However, naval operations, apart from being very expensive, cannot last forever. In addition, one of the technical difficulties is to distinguish between peaceful fishermen and pirates. Even when very close, it is often hard to determine only visually if the boat is a fishing boat or a pirate craft (see Figure 11.3). Also, mother ships resemble larger fishing vessels (see Figure 11.4).

A statement by Allied Maritime Command Headquarters Northwood said that while the Gulf of Aden had seen a drop in piracy over the past year, due in part to the warship patrols from maritime nations, pirate attacks in the Somali Basin, an area the size of Australia, had risen in recent months. UN Secretary General Ban Ki-moon said, "There is simply too much water to patrol, and an almost endless supply of pirates."

Economically seaborne piracy against transport vessels causes estimated worldwide losses of \$13–16 billion per year, particularly in the waters between the Red Sea and Indian Ocean, off the Somali coast, and also in the Strait of Malacca and Singapore.

11.3 The Threat

Modern pirates favor small boats and taking advantage of the small number of crew members on modern cargo vessels. They also use large vessels to supply the smaller attack/boarding vessels. Modern pirates can be successful because a large amount of international commerce occurs via shipping.

Figure 11.3 A suspicious skiff. It is very difficult to distinguish visually a fishing boat from a pirate craft.

Figure 11.4 Mother ships and fishing vessels resemble each other.

Major shipping routes take cargo ships through narrow bodies of water (such as the Gulf of Aden and the Strait of Malacca): these routes make them vulnerable to be overtaken and boarded by small motorboats. Other active areas include the South China Sea and the Niger Delta. As usage increases, many of these ships have to lower cruising speeds to allow for navigation and traffic control, making them prime targets for piracy.

Also, pirates often operate in regions of developing or struggling countries, with smaller vessels and large trade routes. Pirates sometimes evade capture by sailing into waters controlled by their pursuer's enemies. Modern pirates are sometimes linked with organized-crime syndicates, but often they belong to small individual groups.

11.4 Prepare to Repel Boarders!

It is a long time since that command has been issued by the captain of any vessel, as shown in Figure 11.5. Vessels used to carry weapons for just those sorts of eventualities; buccaneers, who preyed on honest merchantmen, knew that if the other vessel's crew won the fight, they probably were not going to collect any loot, but could expect to be swinging from a yardarm in short order. On deck sailors were prepared to lead boarding parties in close action, and to repel enemy boarding parties. The order to "prepare to repel boarders" was issued when a ship was threatened with an enemy assault. The sound of the drum was a signal for the crew to take their stations as in action. The call is beaten in such a manner as to be readily understood and distinguished from all others; everyone on board repairs immediately to his station.

Similarly to what happened in ancient times, when pirates are boarding, it is already too late to put in place the proper strategy in order to prevent the boarding and minimize risk of loss of goods or, even worse, of human lives. Nowadays instead of repelling boarders, the attitude is to prevent pirates from boarding the ship. Of course, merchant ships today want to refrain from violent reactions and avoid carrying armed personnel on board. This means that protection must be mainly passive, increasing surveillance and detection capabilities, while at the same time equipping the ships with the means to make the boarding difficult, as well as providing equipment and networks for launching alerts and calling for support. The longer the time allowed for the ship in engaging in evasive maneuvers and to launch the alarm, the higher is the probability of success against pirates' attacks. Of course, the best result is attained whenever pirates are apprehended before launching their attacks. This is possible whenever surveillance and identification means are put in place.

Figure 11.5 In the nineteenth century, the order "prepare to repel boarders" was issued when a ship was threatened with an enemy assault.

11.5 Modern Solutions for the Antipiracy Issue

A range of options exists for combating maritime piracy. Individual ships have adopted different rudimentary on-board deterrents (i.e., fire hoses and carpet tacks) to repel pirates. More modern solutions must be followed to increase safety and security while sailing in dangerous areas. A rational solution to the problem is based on the presence of appropriate sensors on the ships and a real-time connection with an existing and operating shore headquarters (SHQ). Accordingly, the implementation of a safety route for the ships consists of the presence of two operating segments. There is a ship segment on board that electronically patrols via radar and video the surrounding area and sends this information to the SHQ via radar tracks and video images. At the SHQ there is a shore segment that elaborates the received information, in the frame of its much wider available knowledge of the maritime overall situation, so that warnings to the ship could be provided.

11.5.1 The Ship Segment

The main goal of the ship segment is the early detection of piracy attacks to increase awareness and allow greater response times. In most cases an early warning of half an hour is sufficient to prevent the attack. If pirates are detected at a range of about 5 NM (nautical miles) and recognized at a range of about 2.5 NM, the ship can:

- Actuate evasive maneuvers;
- Be in early contact with SHQ for getting help in advance;
- Prepare dissuasion measures.

The main characteristics that identify pirates from other innocent boaters are: the types of boats (boats about 6 meters long); the composition of crew (three or four persons); weapons (often hidden on the bottom of the boat); and a long (at least 5 meters) ladder. Obviously, the capability to clearly detect weapons (machine guns, rocket propelled grenade launchers, and so forth) is a very desirable feature. Figure 11.6 shows a clear identification of pirates and of their weapons.

11.5.2 The Shore Segment

The main goal of the shore segment is to provide all relevant information to the ships under control. In order to implement this mission, the shore segment must first build up the complete real-time image of maritime traffic (coopera-

Figure 11.6 Pirates are clearly identified by the presence of weapons on board.

tive and noncooperative ships and boats) from all possible sources of information. Then it should assure a wide range of communication services, such as:

- Receive the Ship Security Alert System (SSAS) signals;
- Interface the search and rescue (SAR) with the Maritime Rescue Coordination Centre (MRCC) and the Global Maritime Distress and Safety System (GMDSS);
- Exchange information with all maritime transportation stakeholders;
- Communicate with the ships wherever these might be.

At this stage, the situation awareness of the area where piracy attacks may take place is under control. The counterpiracy situation awareness process can be basically divided into two subprocesses:

- Localize any object in the area;
- Distinguish pirates from all other ships.

Detecting piracy attacks from shore is technically difficult when they are carried out beyond the coastal radar coverage, and, as they are fast and sudden, it is quite unlikely to catch them from satellites that are characterized by relatively long revisiting times. In any case, the warning time might be too short for an efficient reaction. Therefore, one of the main operational goals is to identify their mother ship, as they stay for weeks in the same area and can be detected by satellite or by cooperative ships sailing nearby.

11.6 A Recently Implemented Antipiracy System

Among the possible and suggested up-to-date antipiracy solution, we describe with some detail POMPEIUS, the state-of-the-art antipiracy system developed and experimented by Selex Sistemi Integrati (Selex-SI). This system contrasts modern piracy and safeguards human lives, automatically carrying out the surveillance of the area surrounding the ship, day and night, and detecting possible threats at suitable ranges for starting evasive maneuvers and alerting the closest patrol unit. POMPEIUS is a net-centric architecture system able to answer to the requirements of countering the modern maritime piracy.

According to the typical antipiracy system configuration, POMPEIUS consists of two operating segments: the shore segment and the ship segment. These segments cooperate with each other, but they can also efficiently operate autonomously. When they are integrated, ships send radar and video tracks to the shore headquarters (SHQ), whereas SHQ sends warnings to the ships themselves.

The POMPEIUS ship segment typical architecture consists of the following Selex-SI components:

- A Medusa EO (electro-optical) system;
- A video surveillance unit (VSU) for EO video processing;
- An optional LYRA surveillance radar for classification purposes;
- A multifunctional console for system control and display of radar and EO tracks and video presentation.

This combination of EO, radar, and visual observation, if properly used, can attain a good characterization of the threat (kinematic characteristics of the target, the route followed, and so forth), even if pirates usually tend to disguise their attitude. All these components can be easily installed on commercial ships, with negligible impact. In addition, the modularity of the system allows a wide growth capability (e.g., by installing the components in subsequent phases, yet providing a remarkable feature for the ships).

11.6.1 The Medusa EO System

The Medusa EO system consists of:

- A two-axis motorized director equipped with the electro-optical sensor suite;
- A computer unit.

11.6.1.1 The Medusa Director

The Medusa Director (see Figure 11.7) is characterized by a rugged and rigid structure (in order to minimize sensors vibrations) that allows high-accuracy performances (to keep a narrow beam laser on the target) and high-dynamic performances (maximum speed 2 rad/sec and maximum acceleration 6 rad/sec^2 for each axis).

Azimuth and elevation movements are carried out by means of direct drive motors. This solution achieves the complete elimination of motor/output backlash and high immunity to structural vibration of the vessel, resulting in a better tracking accuracy (rms typically better than 150 mrad), and good quality of video images. Furthermore, a high tracking accuracy allows the use of reduced laser rangefinder beam divergence, with a consequent consistent increase of the tracking range.

The basic EO sensor suite integrated on the Medusa Director consists of a color CCD (Charged Coupled Device) TV camera and an IR camera. A typical IR image, showing how elements at higher temperatures are clearly visible, is depicted in Figure 11.8.

Figure 11.7 The Medusa Director.

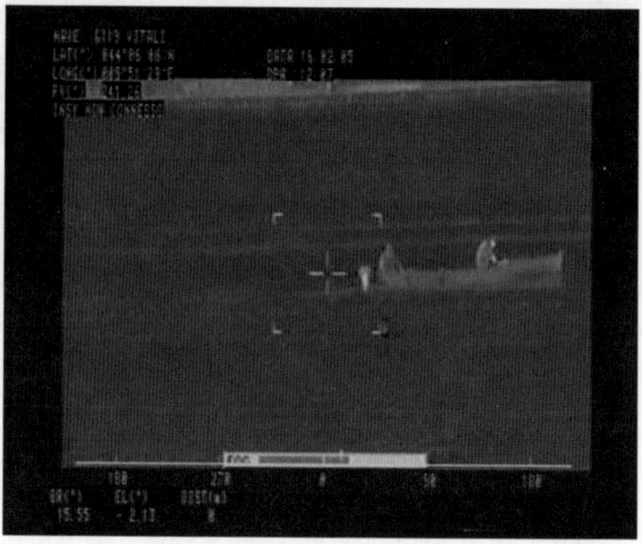

Figure 11.8 Typical IR image: the outboard and even the exposed body parts of the boaters are clearly visible.

TV and IR cameras are equipped with a continuous zoom lens in order to offer excellent situational awareness, while also providing the possibility to zoom in at suspect activities and have a closer look once they are detected.

The narrow field of view is about 1°.

The IR camera is a cooled type, mid-wavelength infrared (MWIR), based on a high resolution (640 × 512 pixels) Cadmium Mercury Telluride (CMT) focal plane array, which offers extremely long-range detection in all weather and light conditions (see Figure 11.9).

An eye-safe laser rangefinder (Safety Class 1) can be added (it is already integrated in the military version for tracking purposes), in order to obtain a three coordinates target position (azimuth, elevation, and range). If correlated with on-board gyrocompass and GPS data, the target geographical position can also be provided.

11.6.1.2 The Medusa Computer Unit

The Medusa Computer unit is based upon state-of-the-art components-off-the-shelf (COTS) processing units and industrial standard I/O cards, hosted in a VERSA Module Eurocard (VME). This is a computer bus standard, originally developed for the Motorola 68000 line of CPUs, but later widely used for many applications and standardized by the IEC rack, complete power supply, and hard real-time operating system. It includes the automatic video tracker; this system processes the video images directly in real time to establish the position

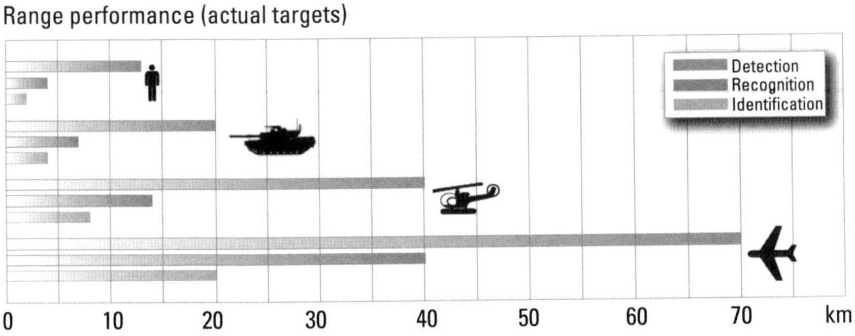

Figure 11.9 POMPEIUS IR camera expected range performances.

of a designated object relative to the sensor boresight. This error is then used to control the director so that the director and attached EO sensors accurately follow the target.

11.6.2 The Video Surveillance Unit

The video surveillance unit allows EO panoramic passive surveillance in order to automatically detect and track small surface targets at a sufficient range to allow for the early warning of piracy attacks. Therefore, electro-optical surveillance effectively complements radar surveillance.

Electro-optical surveillance provides detection and tracking capabilities covering 360° in azimuth and 5° in elevation (selectable within an elevation sector between -20° and +20°). The surveillance scan rate is 0.3 Hz.

Electro-optical surveillance exploits infrared thermal radiation to provide an image of the scenario around the ship. The thermal radiation is acquired in the MWIR infrared band ($3 \div 5$ μm) to allow the acquisition of an optimal image irregardless to what the meteorological conditions would be.

When the electro-optical panoramic surveillance is requested by the operator, the Medusa Director is put in continuous rotation (360° horizontally stabilized) at elevation selected by the operator on the console. A panoramic infrared image of 360° is obtained for display (see Figure 11.10). Each panoramic infrared image is subdivided into five stripes of 72° at a reduced resolution. High-resolution images can be displayed zooming a sector of the panoramic image.

The IR image processing is activated for automatic extraction of targets, with a minimum false alarm rate. Automatically extracted targets are visualized with synthetic marks superimposed to the raw video on the monitor; by means

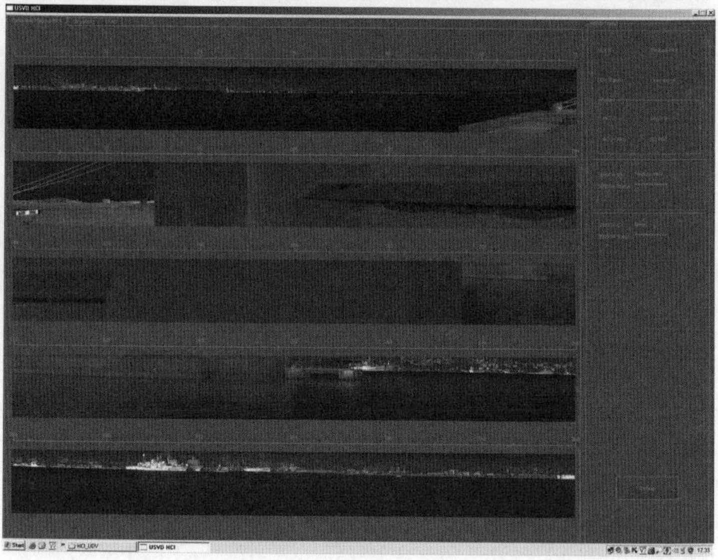

Figure 11.10 EO panoramic presentation. The panoramic infrared image of 360° is subdivided into five stripes of 72°.

of the track ball, the operator can manually select the mark to be investigated (e.g., providing the target angular coordinates to the Medusa Director).

11.6.3 The LYRA Surveillance Radar

LYRA (see Figure 11.11) is a coherent radar that can control and track the frequency and phase characteristics of the transmitted waveform. Wideband

Figure 11.11 The LYRA radar.

frequency modulation (WFM) is used to increase the electromagnetic compatibility and is helpful in reducing interference from other radiating systems.

Frequency diversity is performed using a digital frequency synthesizer to generate multiple waveforms. Pulse compression is performed digitally. This technique allows a very fine tuning of the compressed pulse.

The joint use of frequency modulation and pulse compression allows the utilization of low instantaneous power and long pulses. Thus, electromagnetic power emission is well within the exposure safety limits for operators on board. The LYRA radar's main capabilities are the detection of steady or moving targets and high spatial resolution, which, in turn, provides a rejection of unwanted background echoes and ensures the required system sensitivity. LYRA employs fully solid state technology and very powerful digital processing boards using proprietary state-of-the-art algorithms that offer low transmit peak power, low voltage supply, high compactness, high reliability, and capability of frequency diversity transmission on multiple frequencies. The main system performances are summarized in Table 11.1.

11.6.4 The Multifunctional Console

The multifunctional console (see Figure 11.12) consists of:

Table 11.1 Basic Performances of LYRA Radar

Typical detection range on surface target	48 km
Coverage	Full azimuth coverage or sectorial transmission capability
Nominal resolution	Range 9m Azimuth 0.45°

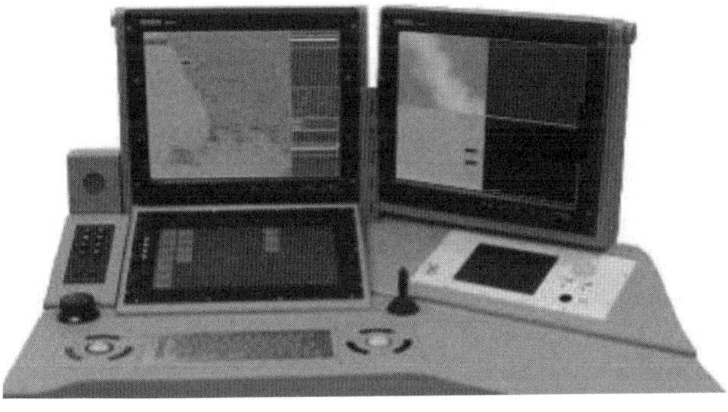

Figure 11.12 The multifunctional console.

- A computer performing human-to-computer interface functions;
- Two high-resolution monitors, in order to display TV and IR videos, EO panoramic presentation, radar presentation, and POMPEIUS parameter information (e.g., system status, fault indication, operational alarms).
- A control desk provided with all the devices needed for the system management (e.g., multifunctional keyboard, track ball, joystick).

11.7 The Validating Experiment

In March 2010 a reduced POMPEIUS ship segment (the LYRA radar was not present) was successfully tested in an operational area on board the ship High Endurance, an oil tanker courtesy of the D'Amico fleet (see Figure 11.13). The test was performed during a normal navigation along the route between Fujairah (UAE) and Sikka (India) (see Figure 11.14).

Integration with on-board navigation radar was realized in order to receive indication of a target detected on video radar. The EO pedestal on-board installation required a support to raise the sensor field of view over nearby obstacles. No interferences with other electronic equipment were experienced. In any case, both system installation and interface with existing equipment installed on board were carried out very easily. The purpose of the experiment was to demonstrate the capability of the POMPEIUS ship segment, in the typical

Figure 11.13 Medusa Director installed on board the tanker of the D'Amico fleet for live testing in the Gulf of Aden.

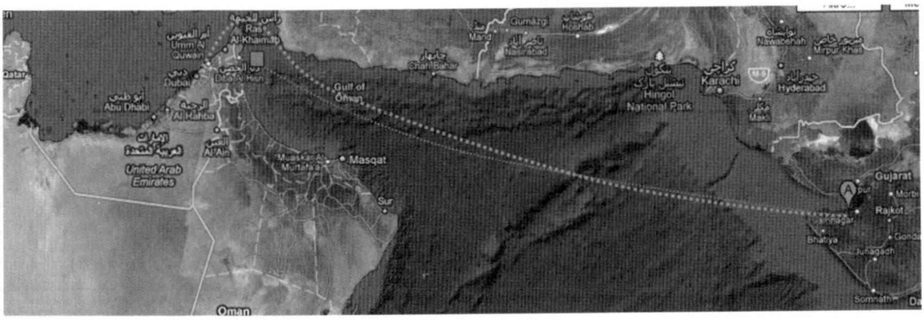

Figure 11.14 Validation experiment operational area.

environment where pirates operate, checking its target detection and recognition capabilities. The integration with navigation radar allowed the display of radar target tracks directly on the POMPEIUS multifunctional console, thus minimizing reaction time in target recognition and optimizing decisions for evasive maneuvers to counter pirate attacks.

During the navigation of about 2,000 NM, several detection and recognition range performance tests on occasional small targets, similar to those utilized for piracy attacks, were carried out. POMPEIUS allowed the recognition of target types (mainly fishing boats) and the identification of crew on board, both day and night. Ambient conditions were those typical of the area, visibility 4–5 NM, temperature 25–35°C, and relative humidity (RH) 70–80%.

The experiment verified that POMPEIUS fully meets the operational needs to prevent pirates from boarding the ships, assuring required target detection, recognition and identification performance ranges.

11.8 Conclusions

Sea piracy is an emerging crime today, with a significant increase rate. In this chapter a modern piracy scenario and a possible counterpiracy approach are presented. These consist of combined ship and shore segments, which allow the identification of the threat and the implementation of the appropriate countermeasures. A system that implements this approach is presented is some detail, and a validating experiment along a normal navigation route between Fujairah (UAE) and Sikka (India) is reported. It is noted that no weapons, as guns or missiles, and weapon-trained people over the ship are needed. However, an appropriate organization over the shore is required to protect and defend the ship itself.

Different types of counterpiracy philosophies might be foreseen. One of the most interesting one is to equip the ship with nonlethal weapons. These are essentially microwave guns that radiate intense microwave beams to interfere and render nonoperative the pirates' boarding ship engines. These techniques have been already experimented over the ground to stop cars, but not yet over the sea to stop vessels; nevertheless, the idea is certainly interesting.

About the Authors

Giorgio Franceschetti has been a full professor at the University Federico II of Naples, an adjunct professor at UCLA, a visiting professor at a number of universities in Europe and the United States, and a lecturer in China and India. He is now the B. Kesseler honorary professor at the University of Trento, Italy, a visiting scientist at Jet Propulsion Laboratory (JPL) in Pasadena, California, and a lecturer at the Space-Tech Master of the University of Delft, the Netherlands. He is author of about 200 papers on peer-reviewed international scientific journals and 10 books. He has received a number of awards, including the Gold Medal from the President of the Italian Republic in 2001, the IEEE Distinguished Achievement Awards from the IEEE Geoscience and Remote Sensing Society in 2007 and the IEEE Antennas and Propagation Society in 2010, and the NASA Group Achievement Award in 2009. Dr. Franceschetti is an IEEE Life Fellow.

Marina Grossi is the CEO of SELEX Sistemi Integrati, a FINMECCANICA company. Dr. Grossi, a university graduate in electronic engineering, has worked in high-responsibility positions in many other FINMECCANICA companies, including Integrati, a FINMECCANICA holding company. Dr. Grossi, a university graduate in electronic engineering, has covered high responsibility positions in many other FINMECCANICA holding companies, such as Alenia Marconi System, MBDA Italia, SMA, Officine Galileo, Alenia Difesa. She was responsible for the Value Creation Project, whose results are available in the book *Value as a Management Model* (2003), and she was the recipient of the Marisa Belisario Award for competitiveness, research, and innovation.

Roberta Cardinali is currently working for SESM, a joint venture between two companies of FINMECCANICA, where she is responsible of several projects funded by the European Community (EU-FP7). In particular, she is the project manager of ATOM (Airport detection and Tracking Of dangerous Materials by passive and active sensors arrays). Dr. Cardinali's main research interests include data fusion, data management, support to decision, and items in the same area. She has expertise in the field of radar system analysis, passive radar, and signal processing.

Nanxi Chen is currently working for Qualcomm Inc. He graduated from Shanghai Jiao Tong University with a bachelor's degree in information security, and then acquired a master's degree in computer science from UCLA in 2010. At UCLA, he worked on the selective dissemination of files and streams to different groups with different qualifications, and certifications in vehicular and tactical ad hoc wireless network scenarios. The research used attribute-based encryption (ABE)–based security and privacy techniques, and was carried out under the guidance of UCLA Professor Mario Gerla.

Fabiola Colone is currently an assistant professor at the DIET Department of the University La Sapienza of Rome. She received a Ph.D. in remote sensing in 2006 from the same university, and has been a visiting scientist in the Electronic and Electrical Engineering Department at the University College of London. Her research interests include passive coherent location, multichannel adaptive signal processing, and space-time adaptive processing with application to monostatic and bistatic radar systems. She is involved in scientific research projects funded by the Italian Space Agency, the Italian Ministry of Research, the Italian industry, and the European community. Her research has been reported in a number of publications on international scientific journals and conference proceedings.

Patrizia Fabbri is currently working for SELEX Sistemi Integrati, a FINMECCANICA company that she joined in 1974. She has been in charge, in the Naval System and Engineering Departments, of design, validation, and testing of integrated radar, artillery, and navigation systems. Dr. Fabbri holds a university degree in mathematics, and was an assistant to the chair of rational mechanics at Genoa University.

Alfonso Farina is the director of engineering of the Large Systems Business Unit at SELEX Sistemi Integrati, a FINMECCANICA company that he joined in 1974. From 1979 to 1985 he was a professor of radar techniques at the University of Naples. His professional activity has been devoted to signal detection, data analysis, image processing and fusion, with very valuable results aimed at the design of radar systems. He is the author of more than 500 technical publications, including three books: *Radar Data Processing* (vols. 1 and 2, John Wiley & Sons 1985–1986, translated also into Russian and Chinese), *Optimised Ra-*

dar Processors (Peter Peregrinus, Ltd., 1987), and *Antenna Based Signal Processing Techniques for Radar Systems* (Artech House, 1992). He has received several technology excellence awards, including the IEEE Dennis J. Picard Gold Medal for Radar Technologies and Applications. Dr. Farina is a Fellow of the IEEE, the Royal Academy of Engineering-UK, IET, and EURASIP.

Pietro Finocchio retired as lieutenant general of the Italian Air Force in 2010. He is currently the president of the Armed Forces Communications and Electronics Association (AFCEA), Rome Chapter, and a member of the Italian Institute of Navigation Presidential Committee. General Finocchio holds a degree in electronic engineering from the University of Naples, and his career has been fully developed within the Italian Air Force organization, from the technical director of the Flight Test Centre up to the general director of TELEDIFE (procurement and contractual top-level agency for ICT and high technology of Italian defense organization). He also has a large experience in the NATO environment, with diffuse participation to panels, delegations, committees, and policy groups.

Mario Gerla is a professor of computer science at UCLA, which he joined in 1976. At UCLA he has designed network protocols including ad hoc wireless clustering, multicast (ODMRP and CODECast), and Internet transport (TCP Westwood). He has led the ONR MINUTEMAN project, designing the next generation scalable airborne Internet for tactical and homeland defense scenarios. He is now leading two advanced wireless network projects under Army and IBM funding: a vehicular testbed for safe navigation, urban sensing and intelligent transport; and personal communications for cooperative, networked medical monitoring. Dr. Gerla is an IEEE Fellow.

Sofia Giompapa is currently working for SELEX-Sistemi Integrati in the Systems and C4I (Command, Control, Communication, Computer and Intelligence) Analysis Group. She has been a consultant at Capraro Technology, in Utica, New York. She also received a bachelor's degree and a master's degree in telecommunication engineering from the University of Pisa, both cum laude, as well as a Ph.D. from the University of Pisa in 2008, doing research in the area of multisensor integrated system for Homeland Security applications.

Antonio Graziano is currently the head of systems and C4I (Command, Control, Communication, Computer and Intelligence) analysis within the Large Systems Business Unit of SELEX-Sistemi Integrati, which he joined in 1990 as system analysis engineer. He received a degree in electronic engineering from the University of Palermo, and has been involved in air defense, surveillance, command and control, and homeland protection systems. He is author of several journal and conference publications, and has also been involved in projects addressing critical infrastructure protection, systems modeling, and interdependency analysis, and information fusion.

Bernhard M. Hämmerli is a professor of networking and information security at the University of Applied Science Lucerne, Switzerland, and University Collage Gjovik, Norway. His main research interests are information security and critical information infrastructure protection (CIIP). He has participated in three EU FP CIIP research programs, the last being on protection and trust in financial infrastructures (www.parsifal-project.eu). He is the editor of the *European CIIP Newsletter*, the president of the Swiss Informatics Society, and the chair of scientific and international affairs in the Information Security Society Switzerland (ISSS). In 2007 he was assigned from Switzerland as a seconded national expert on critical infrastructure protection (CIP), to the Joint Research Centre of the European Commission. He is now chairing the CIP task force at the Centre for European Policy Studies (CEPS) in Brussels. Dr. Hämmerli's consulting corporation, ACRIS, offers services in his expertise area, information, and its critical infrastructures protection.

Uichin Lee is currently working with the Knowledge Service Engineering Department at the Korea Advanced Institute of Science and Technology (KAIST). He received a Ph.D. in computer science from UCLA in 2008, and then worked for Alcatel-Lucent Bell Labs as member of the technical staff until 2010. His research interests include distributed systems and mobile/pervasive computing.

Pierfrancesco Lombardo is currently a professor at the University La Sapienza of Rome. He served in the Official Test Centre of the Italian Air Force, has been an associate with Birmingham University and Defense Research Agency (DRA) in Malvern, United Kingdom, and was a research associate at Syracuse University, New York, in 1996. His research is reported in over 170 publications on international scientific journals and conferences. He received the Best Paper Awards on AES in 2001 and IEEE-GRS in 2003 Transactions. Dr. Lombardo has been an associate editor for *IEEE-AES Transactions* since June 2001, and is a member of the IEEE-AES Radar Systems Panel.

Janos Mathe is currently a graduate student at Vanderbilt University in the Department of Electrical Engineering and Computer Science. While pursuing his Ph.D., he works as a research assistant at the Institute of Software Integrated Systems. He holds an M.Sc. in computer science from the Budapest University of Technology and Economics. His main research area is health information systems with an emphasis on patient management and decision support. He is particularly interested in applying model-based techniques for the representation, simulation, verification, and deployment of computer interpretable medical guidelines.

Roberto Mugavero is a professor of action planning for Homeland Security at the University Tor Vergata of Rome, where he also holds the position of the scientific director of the Master of Science Protection from CBRN Events. He is also the president of the Observatory on CBRN Security and Defense, and is very active in several fire prevention, security, and safety courses, and security, safety, CBRN risk, emergency management, risk management, crime prevention, and technological-industrial risk conferences. As a chemical, bacteriological, radiological, nuclear (CBRN) defense expert, he is a board member of the Italian National Department of Civil Protection, and is an Italian representative in the European Civil Protection Task Force. Dr. Mugavero is also a board member at the Italian Institute for Studies and Research on Security, Civil Protection, and Defense (ISPRO) Institute, and is a volunteer firefighter officer, with the rank of colonel, for the Rome Fire Brigade Board, participating to many national and international exercises and emergency activities.

Soon Y. Oh is currently a postdoctoral fellow at the Network Research Lab at UCLA. He received a Ph.D. in computer science from UCLA in 2009, and has published more than 20 papers in the area of mobile wireless ad hoc networks, aimed for a number of practical applications. He has also implemented and run network coding for ad hoc routing over multiple paths, increasing packet reliability and robustness. At UCLA, he participates in several MANET projects, led by Professor Mario Gerla and supported by DoD, DARPA, ONR, NSF, MURI, ITA, and STTR.

Pompeo Piedimonte is presently serving, with the rank of colonel, in the Command & Control Division of the General Secretariat of Defense and National Armaments Directorate of the Italian Ministry of Defense. He is also the Italian national representative and chairman of the Management Group for Enabling Technology for Advanced Radio in Europe, R&D project. Colonel Piedimonte holds degrees in electronic engineering from the University of Naples and in system engineering from the Italian Defense Staff School of Rome. His career has been fully developed within the Italian Air Force organization, including his assignments as a delegate in NATO meetings and groups.

Leonardo Quattrocchi is currently a senior vice president and the director of public relations and marketing communications at SELEX Sistemi Integrati, which he joined in 1982. He has also worked for other two industrial companies, in charge of marketing and sales, and as the product manager for various lines of products. Dr. Quattrocchi holds a degree in mechanical engineering from the University of Rome. He attended the 18th Session of European Directors of Armament at the École Militaire in Paris, sponsored by the French Centre des Hautes Études de l'Armement, and the Second Session of Istituto Alti Studi per la Difesa (equivalent to a master's degree) of the Italian Ministry of Defense. Dr. Quattrocchi received the Distinguished Cross by the Italian Firefighters Association in 2009 and the Distinguished Service Cross by the President of the Republic of Italy in 2010.

Adarosa Ruffini is a professor at the University of Pisa, where she teaches courses in the area of environment risk, security of production sites, and norm and rules of transportation. She is also professionally active in the area of maritime and international laws. She is the coordinator of the WG (Working Group) Security of Tourist Structures in the TC (Technical Commission) 391, Societal and Citizen Security, of UNI (the Italian Committee for Standardization), the coordinator of WG Modelling of the TC 391 of CEN (European Committee for Standardisation), and an expert in the WG 1, Item Public & Private Partnership, of the TC 223, Societal Security. She has widely published in the area of security rights, risk management, norms, and certifications, and participates in conferences and symposia on security, privacy, and safety.

Shankar Sastry is presently the dean of the College of Engineering and Nippon Electronic Corporation (NEC) Distinguished Professor at the University of California at Berkeley. He was formerly the director of Center for Information Technology Research in the Interest of Society (CITRIS) and the Banatao Institute. He served as the chair of the EECS Department, as the director of the Information Technology office of DARPA at Washington, D.C., and as the director of the Electronics Research Laboratory, an organized research unit on the Berkeley campus, conducting research in computer science and in all aspects of electrical engineering. Professor Sastry is now the president of the Team for Research in Ubiquitous Secure Technology (TRUST), an association of eight top-level U.S. academic institutions aimed at the development of fundamental cyber security science and technology.

Janos Sztipanovits is currently the E. Bronson Ingram Distinguished Professor of Engineering at Vanderbilt University, and is the founding director of the Institute for Software Integrated Systems (ISIS). His research areas are at the intersection of systems and computer science and engineering, including the foundation and applications of model-integrated computing for the design of a wide range of information systems, and additional research contributions about structurally adaptive and autonomous systems, design space exploration, and systems-security codesign technology. He won the National Prize in 1985 and the Golden Ring of the Republic in 1982 in Hungary. Dr. Sztipanovits is an IEEE Fellow and an external member of the Hungarian Academy of Sciences.

Jan Werner is currently a graduate student in the Electrical Engineering and Computer Science Department of Vanderbilt University. He is a research assistant at the Institute of Software Integrated Systems. His main research area is health information systems with an emphasis on privacy-aware design of patient management systems. He received an M.Sc. in computer science from the University of Krakow, Poland.

Stephen B. Wicker is currently a professor of electrical and computer engineering and a member of the graduate fields of computer science, information science, and applied mathematics at Cornell University. He has been awarded three times, by the Cornell College of Engineering (1998, 2009), and the Cornell School of Electrical and Computer Engineering, the Teaching Award (2000), and is the Cornell Principal Investigator for TRUST. He has published four books in the area of coding theory and techniques: *Reed-Solomon Codes and Their Applications* (IEEE Press, 1994), *Error Control Systems for Digital Communication and Storage* (Prentice-Hall, 1995), *Turbo Coding* (Kluwer, 1999), and *Codes, Graphs, and Iterative Decoding* (Kluwer, 2002). Dr. Wicker has also served as an associate editor of IEEE and ACM Transactions. He is an IEEE Fellow.

Index

Access gateway (GTWA)
defined, 110
failure, 111, 112
See also Critical infrastructures (CI)
Accountability, 212
Account accreditation, 71
Air France Flight 8669, 197
Air-India Flight 182, 197
Airport security, 173–93
ATOM project, 174, 175, 181
carry-on luggage screening, 177
Checkpoint of the Future, 178–79
costs, 180
dangerous materials detection, 173–93
distributed security system scenario, 182, 183
distributed security system sensors, 184–88
explosive trace detectors (ETDs), 177
improved explosive devices (IEDs), 177
indoor/outdoor noncooperative people tracking, 188–92
innovative approach to, 180–84
off-limits area, 176
passenger screening, 177–78
process perception, 179
process time, 179
public area, 176
quality evaluation, 179–80
screening quality, 179
standard security solution/issues, 176–80
sterile area, 176
Airport security failures, 197–99
Air France Flight 8669, 197
Air-India Flight 182, 197
Cagliari Airport attack attempt, 198
Gatwick Airport attack attempt, 198
Glasgow Airport attack, 199
Iberia office bombing, 198
Lod Airport massacre, 197
Los Angeles Airport (LAX) attack, 198
Madrid Airport bombing, 199
Northwest Flight 253, 199
Pan Am Flight 103, 197
Rome airport attack, 197
September 11, 2001, 197–98
Sibir Airlines flight, 198
Vienna airport attack, 197
Anonymization, 79
Antipiracy system, 264–70
LYRA surveillance radar, 268–69
Medusa EO system, 264–67
multifunctional console, 269–70
POMPEIUS, 264–70
validating experiment, 270–71
video surveillance unit, 267–68
See also Pirates and piracy
Archetypal approach, 211–15
accountability, 212
fairness, 212–13
implementation steps, 213
ISO standard, 211
origins, 211

Archetypal approach (continued)
participation, 213
principle compliance, 213–14
structural components, 214
transparency, 213
See also Privacy; Security
ATOM project, 174, 175
area under surveillance outside gate, 181
security screening, 181
Attribute-based encryption (ABE), 156–58
Attribute update, 160
Authentication, in financial infrastructures, 18
Autonomic network management, 89–90

Battlefield Target Identification Device (BTID), 52
Behavior extraction algorithm, 232, 233
Berkeley/Intel mote, 217
Berlin Air Safety Center (BASC), 206
Beslan hostage crisis, 3
Beta-particle generated EMF, 219
Beta-voltaic power cell, 220
Bilayer lipid membrane, 219
Black box, 87
Blackouts
cascading failures, 100
cascading model, 102–3
dynamic of, 100

C2W. *See* Command and Control Wind Tunnel
Cagliari Airport attack attempt, 198
CareNet
defined, 33
focus, 34
graphical representation, 33
Carry-on luggage screening, 177
CENS' Urban Sensing project, 145
C(I)P, 65–67
Cipher-policy ABE (CP-ABE), 156–58
Civil military cooperation (CIMIC), 45
Closed-circuit television (CCTV)
cameras, 136–37
disadvantages, 140
User Group, 137
Code analysis, 20
Command, Control, Communications, Computers, Intelligence, Surveillance, Target Acquisition, and Reconnaissance (C4ISTAR), 47, 51, 52
Command and Control Wind Tunnel (C2W)
defined, 30
simulation components, 30–31
software infrastructure, 31
Complex systems
adaptive, 86
autonomic network management and, 89–90
challenge of, 86–87
defined, 85–86
emerging phenomena in, 85
high-voltage electric transmission grid study case, 100–105
interdependence analysis in, 105–14
network traffic congestion study case, 90–100
nonadaptive, 86
operational factors, 88
performance evaluation, 89
performance metrics, 89
polynomial complexity, 87
science of complexity and, 88–89
similar components, 86
simulation structures, 88
study cases, 90–114
systems engineering, 87–90
Computer emergency response team (CERT), 76
Computer security incident response teams (CSIRT), 76
Control systems
CI and, 61–63
real-time computing (RTC), 62
Cornell processor, 221
CRAWDAD, 168
Critical financial infrastructure (CFI), 71–76
Critical information infrastructure protection (CIIP), 60–61, 107
Critical infrastructure protection (CIP), 60, 107
Critical infrastructures (CI), 58
access gateway, 110–12
airports, 108
applications, 218
communications, 107
control systems and, 61–63
electrical grid, 107

examples of, 105
graphical area, 109–10
harbors, 108
highway network, 108
ICT and, 58–61
interdependence analysis in, 105–14
interdependencies as new vulnerability, 106–7
models of, 107–8
privacy-aware design, 217–35
protection, 106
railway network, 108
sectors, 59
simulation process scheme, 110
simulator model, 108–9
sociotechnological scenario, 105–6
study case, 105–14
study case conclusions, 112–14
study case details, 109–12
urban area, 108
work within, 59–60
Critical Infrastructure Simulation by Interdependent Agents (CISIA), 108–9
defined, 108
dependences and interdependencies, 108
entity evolution, 109
investigations, 113–14
quantities in, 113
Critical Infrastructure Warning Information Network (CIWIN), 66
C-VeT, 167–72
architecture, 167–70
Census platform, 167
defined, 167
infrastructure, 169
infrastructure illustration, 170
mesh node, 169
mobile node, 168, 169
mobility traces database, 168
MobiMESH, 170–72
provisions, 167–70
vision, 170
Web interface, 168
Cyber-Defense Technology Experimental Research Test Bed (DETER), 28–30
capabilities, 28–29
challenges, 29–30
defined, 28
Cyber security research, 12

DARPA Information Science and Technology (ISAT) format, 13
Data collection
consent for data collection, 229
full disclosure of, 228–29
personal, minimizing, 230
Data Distribution Service (DDS), 19
Datalog, 125
Data retention, minimizing, 230–31
DC power flow
equations, 101
optimal, problem, 100–102
Decryption, 160
Defense Information Infrastructure (DII), 46, 51, 53
Delay/Fault-Tolerant Mobile Sensor Network (DFT-MSN), 145
Demand response case study, 231–35
architecture illustration, 235
power consumption data, 231–33
privacy-ware system, 233–35
Demand response SCADA test bed, 32–33
DETER. *See* Cyber-Defense Technology Experimental Research Test Bed
Digital Rights Management (DRM), 28
Distributed security system
performance, 183
required characteristics, 184–85
scenario, 182
scope, 182
sensors for, 184–88
structure, 183
See also Airport security
Domain Specific Modeling Languages (DSMLs), 23
Dynamic attribute-based encryption (DABE), 155–56
attribute prevention, 161
benefits of, 160–62
components, 158
CP-ABE versus, 158
in dynamic environments, 158–60
future, 162
in HD applications, 162–63
in homeland defense applications, 162–63
trust, 162
Dynamic Security Skins, 15

E-banking
countermeasures against fraud, 69–71
counterreaction, 70
denial of money transfer, 70
general purpose computers for, 67–69
information security challenge, 67–71
repression, 70
Encryption, 159–60
European Convention on Human Rights (ECHR), 209
European Network Information Security Agency (ENISA), 77
European Programme for Critical Infrastructure Protection (EPCIP), 66
Experimental Social Science Laboratory (Xlab), 31
Explosive trace detectors (ETDs), 177

Fairness, 212–13
Financial infrastructures, 14–15
applicable network security for, 19
authentication and, 18
broader needs, 18
code analysis, monitoring, and malware detection, 20
components, 14
core system support for, 18–19
critical (CFI), 71–76
financial transactions, 14
Internet operations and, 18
malware and, 18
public policy, decision making, and risk management, 20
tailsearch in secure Web security, 19–20
TRUST research, 17–20
First in first out (FIFO) rule, 94
Freedom, protection of, 204

Gatwick Airport attack attempt, 198
Geographic Information System (GIS) techniques, 109
Glasgow Airport attack, 199
Global Environment for Network Innovation (GENI) program, 163
Globalization, 65

Healthcare organizations (HCOs)
leveraging of HISs, 119–20
migration, 119
Health information systems (HISs)
advanced, development of, 130
ethical and legal analysis, 127–28
evaluation of, 122
experimental implementations, 122
healthcare delivery complexities, 131
MIC principles, 129
model-based design of, 128–29
model-based experiment integration, 130
national scale deployment, 131
notational structure for development, 121
patient-centered clinical information management processes, 122–24
policy composition, 125–26
policy languages, 125–26
policy role in design, 128
progress evaluation, 127
SECTET, 130
Health infrastructures, 15–16
HIS project, 16, 21, 23–24
home-based health monitoring, 24–25
personal health records (PHRs), 15
privacy modeling and analysis, 21–22
TRUST research, 21–25
Health Insurance Portability and Accountability Act (HIPPA)
Privacy Rule, 22, 122, 127
Security Rule, 122, 123
Heterogeneous traffic congestion, 95–97
High-voltage electric transmission grid study case, 100–105
blackout dynamic, 100
cascading blackout model, 102–3
conclusions, 105
optimal DC power flow problem, 100–102
results, 103–5
Home-based health monitoring, 24–25
Homeland defense (HD)
central office, 162, 163
DABE in, 156, 162–63
Homeland protection, 45
Homeland security
armed forces contribution to, 45
information system challenges and, 57–82
Homeland security scenario
last step, 9–10
network traffic congestion in, 97–99

new, unveiling, 1–6
new rules and operational avenues, 6–9
new vision of, 1–10
Home monitoring test bed, 33–34
Homogeneous traffic congestion, 93–95

Iberia office bombing, 198
Improved explosive devices (IEDs), 177
Information and communication technologies (ICT)
blurring borders, 64
component access, 57
critical infrastructure and, 58–61
generation length, 82
globalization, 65
as part of sectorial infrastructure, 65
Information market, privacy and, 227
Information sharing
key challenges, 76–79
vulnerability minimization, 78–79
Information Sharing and Analysis Center (ISAC), 76
Information technology (IT) systems, 90
Instruction memory (IMEM), 220
Intelligent electronic devices (IEDs), 32
Internet operations, in financial infrastructures, 18
Internet/phone cable cutting, 5
Interpolation, 93
Intrusion Detection System (IDS), 32
Inverse power law behavior, 104
Inverse SAR (ISAR), 186
IR cameras, 265–67
IrisNet, 145
Italian civil defense organization, 242
Italian civil protection organization, 243
Italian G8 Summit, 238
air security, 250–51
armed forces in security activities, 251–52
armed forces operation room, 254
background, 246–48
command and control security system, 252, 253
conclusions, 254–55
contingency plans, 252
coordination table, 252
earthquake and, 247–48
international operations room, 254
joint operation room, 252–53
land security, 250
as major event, 238–39, 248
organization, 249–52
organizational structure, 249
participants, 248–49
planning, 246–54
prospects and problems, 239–40
special security activities, 251
system criticality, 244–46
technical management center, 253–54
Italian national protection system, 240–44
activities of, 240
national emergency management system, 241–44
planning scenarios, 240
recovery and consequence management, 244
risk assessment, 240
security tasks, 240–41
Italian Privacy Code, 210
Italy
Center of Excellence in Modeling and Simulation, 54
MCCIS, 52
NEC implementation, 51–54
SIACCON, 52
SICCAM, 52
SICOT, 53
SICRAL 1/SICRAL 2, 53
IT infrastructure, 50

Joint Research Center (JRC), 77

Katz v. United States, 225
Kyllo v. United States, 226

Lod Airport massacre, 197
Logic of Privacy and Utility (LPU), 22
London Bombing, 3
Los Angeles Airport (LAX) attack, 198
LYRA surveillance radar, 268–69

Madrid Airport bombing, 199
Major events, 238–49
Malware
detection, 20
in financial infrastructures, 18
Medusa EO system, 264–67
elements, 264–65

Medusa EO system (continued)
  Medusa Computer unit, 266–67
  Medusa Director, 265–66
MELANI project, 77, 78, 79
MetroSense, 145
Millimeter waves, 185
Millimeter-wave sensors, 186
Minor operational level (MLO), 52
MobEyes, 141–54
  application developers, 147
  architecture, 146–48
  Data Processor (MDP), 146, 147
  data validation, 149–52
  defined, 142
  Diffusion/Harvesting Processor (MDHP), 146
  effectiveness, 152
  harvesting agents, 144
  Manhattan (MAN), 148, 150, 151, 152
  opportunistic sensor networking and, 144–46
  passively harvested summaries, 150, 151
  prototype, 146–47
  Random WayPoint (RWP), 148, 150, 151, 152
  Sensor Interface (MSI), 146
  sensor node architecture, 146
  tracking suspects with, 152–54
  trade-off configuration, 153
  urban surveillance examples, 148–54
  vehicle sensor platforms and, 144
  vehicular sensor networks and, 142–44
  Westwood scenario, 148–49
  wireless-enabled vehicles and, 143
Mobile sensor platforms, 135–64
  emergence of, 140
  introduction to, 135–41
  MobEyes, 141–54
  urban surveillance, 135–36
  video camera surveillance, 136
Mobile TAN (MTAN), 70
Mobility trend, 63–64
MobiMESH, 170–72
  access network, 171, 172
  ad hoc extension, 171
  characteristics, 170
  defined, 170
  mesh backbone, 171
  mesh routers, 172
  network architecture, 171
  wireless mesh routers, 171
  See also C-VeT
Modeling and simulation (M&S), 87
Model-Integrated Computing (MIC), 129
Model of Integrated Clinical Information Systems (MICIS), 16, 24
Model of integrated computing (MIC), 23
Multifunctional console, 269–70
Multiscenario crisis events, 237–55
Murrach Federal Building Attack, 3

National defense, 45
National emergency management system, 241–44
NATO network-enabled capability (NNEC), 49
NCW. *See* Net-centric warfare
Net-centric warfare (NCW)
  concept, 48
  doctrine, 48–51
  effectiveness, 49
  traditional criteria and, 50
Network-Centric Assessment Tool (NCATTM), 46
Network Centric Operations Industrial Consortium (NCOIC), 46
Network-enabled capability (NEC), 45–48, 49
  implementation in Italy, 51–54
  transformation challenges, 54
Network security, for financial infrastructure, 19
Network traffic congestion study case, 90–100
  conclusions, 97–100
  congestion phenomena, 93–97
  defined, 90–91
  engineering fields, 91
  heterogeneous traffic case, 95–97
  in homeland security scenarios, 97–99
  homogeneous traffic case, 93–95
  interpolation, 93
  mapping, 93
  networks in, 92
  network topological properties, 91–93
  simulation parameters, 94, 95, 96, 97
  See also Complex systems
New Generation Identification Friend or Foe (NGIFF), 52
New world scenario, 43–44

Next generation SCADA (NG-SCADA), 26–27, 220–21
Northwest Flight 253, 199

Object Constraint Language (OCL), 24
Open model of assurance, 209–11
OPEN WRT, 167
Operational avenues, new, 6–9
Opportunistic sensor networking, 144–46
Optimal DC power flow problem, 100–102

Pan Am Flight 103, 197
Parsifal
  documentations, 75
  ideas and paradigms, 73
  project overview, 72–76
  purpose of, 71
  recommendation dependencies, 75
  recommendations and research directions, 74
  topics, 72–73
PARSIFAL project, 58
Participation, 213
Passenger screening, 177–78
Passive bistatic radar (PBR), 189, 191
Passive sensors, 185
Patient-centered clinical information management processes, 122–24
  evaluation, 130–31
  patient management systems, 124
  patient portals, 123–24
Patient management systems, 124
Patient portals, 123–24
People tracking, 188–92
  concept challenges, 189
  PBR, 189, 191
  subsystem, 188
  WiFi technology, 190–91
  See also Airport security
Personal health records (PHRs), 15
Physical infrastructures, 16–17
  demand response technology, 27
  NG-SCADA, 26–27
  privacy and monitoring policy, 27–28
  privacy issues, 17
  SCADA, 16, 25–27
  scalable and secure routing, 26
  secure control systems, 27
  security requirements, 16–17
  TRUST research in, 25–28
PICOs, 81
Pirates and piracy, 258–59
  antipiracy system, 264–70
  conclusions, 271–72
  identification, 263
  modern solutions to, 262–64
  "prepare to repel boarders" and, 261
  ship segment, 262
  shore segment, 262–64
  statistics, 258–59
  target identification of, 258
  threat of, 259–60
Platform-based design (PBD), 23
Policy composition, 125–26
Policy languages, 125–26
POMPEIUS
  architecture components, 264
  IR camera expected range performances, 267
  LYRA surveillance radar, 268–69
  Medusa EO system, 264–67
  multifunctional console, 269–70
  validating experiment, 270–71
  video surveillance unit, 267–68
  See also Antipiracy system
Power consumption data
  consumer-specific, 234
  privacy-aware design, 233–35
  privacy issue, 231–33
PrimeLife, 81
Privacy
  advanced sensing system and, 231–35
  archetypal approach, 211–15
  control suggestions, 81
  defined, 203
  definitions and the law, 224–26
  as fundamental right, 203
  market for information and, 227
  as meaningful value, 205
  need for protection, 202, 203
  in patient management systems, 124
  protection with privacy-aware design, 228–31
  regulation concepts, 80
  rights, erosion of, 204
  security dilemma, 79–82
  security versus, 201–15
Privacy-aware design, 217–35
  consent for data collection, 229
  for critical infrastructures, 217–35

Privacy-aware design (continued)
data identification minimization, 230
data retention minimization, 230–31
demand response case study, 231–35
framework, 228
full disclosure of data collection, 228–29
introduction to, 217–18
personal data collection minimization, 230
privacy protection with, 228–31
SCADA example, 220–24
sensor networking technologies, 218–20
Private key generation, 160
Probability density function (PDF), 103, 104
Programmable logic controllers (PLCs), 32
Protocol Composition Logic (PCL), 19
ProVision Body Scanning Checkpoint Security System, 186
Public policy, 20
Pwd Hash, 15

Qualities-of-service (QoS) properties, 19

Radio frequency identification (RFID) devices, 188
Random distribution, 102
Real-time computing (RTC), 62
Regulatory framework, 209–11
Remote terminal units (RTUs), 221, 223
Residential power monitoring, 224–26
Risk management, 20
Role-based access control (RBAC), 123
Rome airport attack, 197
Rules, new, 6–9

Safety
as basic human right, 204
defined, 205
integrated security management, 205–7
pursuit of, 204
See also Privacy; Security
Sanitarization, 79
Screening
ATOM project, 181
carry-on luggage, 177
in new security role interpretation, 207
passenger, 177–78
quality, 179
See also Airport security

SECTET, 130
Security, 205
airport, 173–93
archetypal approach, 211–15
as basic human right, 204
control instruments, 80
defined, 205–6
freedom-maintaining controls, 81
integrated safety management, 205–7
Italian G8 Summit, 250–52
as meaningful value, 205
need for protection, 202
network, 19
in patient management systems, 124
privacy versus, 201–15
role, new interpretation on, 207–9
vehicular networks, 154–55
Security Breach Notification Acts, 128
Security-privacy dilemma, 79–82
Security Scan sensors, 186
SenseWeb, 145
Sensor networking technologies, 218–20
Sensors
for distributed security system, 184–88
millimeter-wave, 186
passive, 185
UWB, 187, 193
W-band, 187
Sepsis treatment enhanced through electronic protocolization (STEEP), 24, 124
September 11, 2001, 197–98
Service-oriented architecture (SOA), 23, 24, 130
applying, 130
in combination with MIC and PBD, 23
in e-commerce, 130
Setup, 159
Shared experimental facilities, 28–34
Sibir Airlines flight, 198
Smith v. Maryland, 225
Software-defined radio (SDR), 52
Stuxnet computer worm, 5–6
Supervisory control and data acquisition (SCADA), 13, 218
architecture, 222
control schemes, 26
corporate network connections, 64
defined, 62, 221
demand response test bed, 32–33

devices, 26
digital control systems (DCS) for, 25
example application development, 220–24
Internet attacks and, 63
network challenges, 25–26
next generation (NG-SCADA), 26–27
physical infrastructures and, 16
secure control of network, 27
test beds, 31–33
Surveillance
urban, 135–36, 148–54
in vehicular networks, 154–63
video camera, 136–40
SWIFT, 81
SWIM, 145
Synthetic aperture radar (SAR), 186
System criticalities, 244–46
national emergency response systems and, 245
systematic approach and, 245–46
types of, 244–45
Systems, Capabilities, Operations, Programs, and Enterprises (SCOPE), 46
Systems engineering, new approach to, 87–90

Team for Research in Ubiquitous Security Technology (TRUST), 7–9, 16
architectures, 16
Center, 13
competences, 9
creation of, 11
defined, 8
design, 16
examples of, 119
financial infrastructures, 14–15
functioning of, 11–12
grand challenge research areas, 13
health infrastructures, 15–16
information flows, 120
as integrated structure, 9
introduction to, 11–12
legacy of, 13
leveraging, 119–20
model-based design of, 119–31
national goals, 21
NSF Science and Technology Center, 220
objectives, 21
operational avenues, new, 120
physical infrastructures, 16–17
purpose of, 11
research overview, 12–17
SCADA testbed, 223
in secure financial infrastructure research, 17–20
in secure health infrastructure research, 21–25
in secure physical structure research, 25–28
shared experimental facilities, 28–34
strategic planning process, 8
success areas, 8, 12
trusted, architectures for, 21, 23–24
trustworthy HIS development of, 120–22
Tennessee-Eastman chemical plant, 224
Terahertz spectroscopy, 187
Terrorist attacks, 5–6
Beslan hostage crisis, 3
graphical representation, 4
Internet/phone cable cutting, 5
London Bombing, 3
Murrach Federal Building Attack, 3
Stuxnet computer worm, 5–6
Tokyo Chemical Sarin Attack, 2–3
U.S. electric grid penetration, 4–5
Think-Trust, 81
Tokyo Chemical Sarin Attack, 2–3
Top-down process, 214
Transparency, 213
Trojan horses, 68
TRUST. *See* Team for Research in Ubiquitous Security Technology
2-D cross-correlation function (2D-CCF), 190, 192

Ultrawideband (UWB) sensors, 187, 193
United States v. Miller, 225–26
Urban defense, with mobile sensor platforms, 135–64
Urban Sensing project, 145
Urban surveillance, 135–36
maximum uncovered intervals per node, 153
MobEyes effectiveness in supporting, 152
security requirements, 155–56
U.S. electric grid penetration, 4–5

Vehicle sensor platforms, 144

Vehicular ad hoc networks (VANETs), 142, 145
Vehicular networks
attack protection, 155
deployment, 154
security, 154–55
selective surveillance in, 154–63
Vehicular sensor networks (VSNs), 142–44
custom design protocols, 144
defined, 142
in emergency recovery, 143
proactive urban monitoring, 143
VERSA Module Eurocard (VME), 266
Video cameras, 163
Video camera surveillance, 136–40
CCTV, 136–40
fixed, limitations of, 140–41
maps, 138, 139
networks, 137
public support for, 138–39
Video surveillance unit, 267–68
Vienna airport attack, 197
Virtualization, 65

W-band sensors, 187
Web security, for financial infrastructure, 19–20
Westwood MobEyes scenario, 148–49
Wideband frequency modulation (WFM), 268–69
WiFi technology, 190–91

Xlab, 31

ZebraNet, 145